ELECTRONICS

ELECTRONICS

THE LIFE STORY
OF A TECHNOLOGY

David L. Morton Jr. and Joseph Gabriel

GREENWOOD TECHNOGRAPHIES

GREENWOOD PRESS
Westport, Connecticut • London

Library of Congress Cataloging-in-Publication Data

Morton, David, 1964–
    Electronics: the life story of a technology / David L. Morton Jr. and Joseph Gabriel.
      p.   cm.—(Greenwood technographies, ISSN 1549–7321)
    Includes bibliographical references and index.
    ISBN 0–313–33247–9 (alk. paper)
      1. Electronics—History.   I. Gabriel, Joseph.   II. Title.
TK7809.M67   2004
621.381'09—dc22   2004054390

British Library Cataloguing in Publication Data is available.

Library of Congress Catalog Card Number: 2004054390
ISBN: 0–313–33247–9
ISSN: 1549–7321

First published in 2004

Greenwood Press, 88 Post Road West, Westport, CT 06881
An imprint of Greenwood Publishing Group, Inc.
www.greenwood.com

Printed in the United States of America

∞™

The paper used in this book complies with the
Permanent Paper Standard issued by the National
Information Standards Organization (Z39.48–1984).

10 9 8 7 6 5 4 3 2 1

Every reasonable effort has been made to trace the owners of copyright materials in this
book, but in some instances this has proven impossible. The author and publisher will be
glad to receive information leading to more complete acknowledgments in subsequent
printings of the book and in the meantime extend their apologies for any omissions.

Contents

Series Foreword

In today's world, technology plays an integral role in the daily life of people of all ages. It affects where we live, how we work, how we interact with each other, and what we aspire to accomplish. To help students and the general public better understand how technology and society interact, Greenwood has developed *Greenwood Technographies*, a series of short, accessible books that trace the histories of these technologies while documenting *how* these technologies have become so vital to our lives.

Each volume of the *Greenwood Technographies* series tells the biography or "life story" of a particularly important technology. Each life story traces the technology, from its "ancestors" (or antecedent technologies), through its early years (either its invention or development) and rise to prominence, to its final decline, obsolescence, or ubiquity. Just as a good biography combines an analysis of an individual's personal life with a description of the subject's impact on the broader world, each volume in the *Greenwood Technographies* series combines a discussion of technical developments with a description of the technology's effect on the broader fabric of society and culture—and vice versa. The technologies covered in the series run the gamut from those that have been around for centuries—firearms and the printed book, for example—to recent inventions that have rapidly taken over the modern world, such as electronics and the computer.

While the emphasis is on a factual discussion of the development of the technology, these books are also fun to read. The history of technology is full of fascinating tales that both entertain and illuminate. The authors—all experts in their fields—make the life story of technology come alive, while also providing readers with a profound understanding of the relationship of science, technology, and society.

Preface

This book has its origins in preparations for the fiftieth anniversary celebration of the Institute of Electrical and Electronics Engineers (IEEE) Electron Device Society (EDS) in 2002. This interest group of the IEEE sponsored the collection of oral history interviews of some of their prominent members. Plans for a book-length history of electron devices were canceled, and instead the society sponsored a small traveling exhibit and a commemorative pamphlet. That left the authors with a great pile of notes for a book on the subject, but no publisher. Nonetheless, members of the EDS, including history committee chair Craig Casey, provided valuable guidance on the initial phase of the project. The support of the IEEE History Center at Rutgers University was also crucial in the completion of this book. The staffs of the public relations departments of Intel and Lucent Corporations, particularly Ed Eckert and Richard Teplistky of Lucent, generously provided photographs and granted permission to publish them.

Timeline

1904   Fleming valve (vacuum tube diode) invented.

1907   Semiconductor diode (point contact type) invented.

Triode (De Forest Audion) invented.

1947   Point contact transistor invented.

1951   Junction transistor invented.

1954   Junction diode invented (used as a solar cell).

Maser invented.

1956   Nuvistor invented.

1959   Integrated circuit invented.

Planar transistor invented.

1960   Laser demonstrated.

Resistor-transistor logic chips introduced.

Metal-oxide-semiconductor field-effect transistor (MOSFET) invented.

1962   Diode-transistor logic (DTL) chips announced.

Emitter-coupled logic (ECL) chips introduced.

	Red-light-emitting diode demonstrated.
	Semiconductor laser demonstrated.
1963	Complementary metal oxide semiconductor (CMOS) chips introduced.
	Transistor-transistor logic (TTL) chips commercialized.
1968	Sony Trinitron cathode ray tube (CRT) introduced.
	CMOS logic chips introduced.
circa 1970	Large-scale integration announced.
1970	320-bit random-access memory (RAM) chip introduced.
circa 1971	Alphanumeric light-emitting diode (LED) displays introduced.
1971	Charge-coupled device (CCD) introduced.
	Liquid crystal display (LCD) introduced.
	Intel 4004 microprocessor introduced.
1972	1-Kbit (Kilobit) RAM memory chip commercialized.
1974	Intel 8080 microprocessor introduced.
	Motorola 6800 microprocessor introduced.
1975	4-Kbit RAM memory commercialized.
1976	16-Kbit RAM memory commercialized.
	Zilog Z-80 introduced.
1978	Intel 8086 microprocessor introduced.
1979	Intel 8088 microprocessor introduced.
1980	Very large-scale integration (VLSI) announced.
1982	256-Kbit dynamic random-access memory (DRAM) commercialized.
	Intel 80286 microprocessor introduced.
1985	Intel 80386 microprocessor introduced.
	Flash memory commercialized.
1986	1-Mbit (megabit) DRAM commercialized.
1988	4-Mbit DRAM commercialized.
1991	16-Mbit DRAM commercialized.
1992	Intel Pentium announced.

1994   64-Mbit DRAM commercialized.

1998   256-Mbit DRAM commercialized.

2000   Intel Pentium IV microprocessor commercialized.

1

The Origins of Electronics, 1900–1950

◆

INTRODUCTION

Today's computers, televisions, telephones, and all other electronic systems rely on circuits, which are the paths that electricity takes through various electrical components in order to perform some kind of useful task. A simple circuit, for example, might consist of a battery with its positive terminal connected by a wire to one end of a light bulb filament, then a wire leading from the filament's other end back to the battery. It is beyond the scope of this work to explain in detail the physics of electrons or why they cause the filament of a light bulb to emit heat and light. The authors assume that readers will have a little knowledge of physics and electrical circuits, although we believe that only a modicum of such knowledge is necessary to appreciate this book. In fact, the simplified technical explanations presented in this book, while they may suffice in a work of history, will not satisfy readers with a strong background in physics or electronics, nor are they meant to.

GUTS

The present work is primarily a history of the "guts" of modern electronic systems, specifically, the things that electrical engineers refer to as the "active" electronic devices that make up circuits. Defining "active" is, however, something of a problem. Wires, capacitors, and resistors are examples of "passive" electrical devices. While they are essential in virtually any circuit, these devices do not necessarily transform or redirect the current that flows through them. They are in some sense passive because current flows through them. Active devices, however, can perform more profound changes on a current or a voltage, such as amplifying it or switching it on and off. The division between active and passive is admittedly imperfect, because some devices such as microprocessors contain both active and passive elements, and some basic categories of devices, such as diodes, are arguably somewhere between active and passive.

It may at first seem arbitrary to exclude passive devices and minimize the discussion of whole circuits, given that the devices themselves are as meaningless outside the context of a circuit as would be a history of tires without a discussion of the automobile. There are, however, justifications for a focus on active devices almost exclusively. It is primarily active devices, such as the cathode ray tube (CRT), the transistor, the laser, and the microchip that have stimulated public interest and have become household words. Their significance is suggested by the fact that the inventors of these devices have in several cases received Nobel Prizes or other major international awards. Part of the reason for this is that these active devices form the hearts and brains of important technological systems. They are important not only for scientific and technical reasons, but also for symbolic or cultural reasons, just as human hearts and brains are. We believe that the growing awareness of the importance of these devices, contrasted with the general lack of knowledge about them, is a compelling reason to study their origins and development.

EARLY HISTORY

Contemporary distinctions in engineering between active and passive devices, and even the term "electronics," were imposed on the technology retroactively, long after the first active electronic devices were actually invented. Looking back, engineers of the 1920s used the term "electronics" in part to distinguish the new technology of radio from the older technologies of electric power, lighting, motors, the telephone, and the telegraph. Before there was electronics, there was simply electricity. The beginning of the development of electrical technology in the 1700s was the discovery of the mechanical means to build up charges of static electricity and transmit them along wires, or store them in crude "accumulators" called Leyden jars (precursors of the modern capacitor). Around 1800, Italian physicist and inventor Alessandro Volta discovered the electric battery, a device he called the pile, and later in the century other inventors found ways to build mechanical electric generators, which provided sources of electricity. Using electricity for practical lighting, heating, or mechanical work (using motors) had all been accomplished before 1850, although it was not very successfully commercialized. Electricity in 1850 was still largely a scientific phenomenon but appeared to be on the verge of finding a market.

Most of the devices covered in this book fall under the category of "electronic" rather than simply electrical, and the distinction is important. Early twentieth-century radio engineers began referring to devices as "electronic" if their function depended on the flow of electrons through free space. At the time, they were thinking mainly of the new class of devices known as vacuum tubes (discussed later in this chapter), which were then coming into wide use as detectors of radio signals, generators of radio-frequency waves, and amplifiers. The progenitor of twentieth-century vacuum tubes was the glow tube, a sealed glass tube from which most of the air had been pumped out and replaced with some other kind of gas. If two wire electrodes were inserted into the ends of the tube (this required that the openings in the tube were tightly sealed, but the seal was rarely perfect) and a generator or other source of current was attached, the gas inside the tube would ionize and set up a conducting path for electrons from the negative electrode (or cathode) to the positive electrode (or anode), which along the way would cause the gas to glow. At the time, however, the mechanism for this phenomenon was not known. The existence of the electron had not yet been proven, and electricity was thought of as a sort of invisible fluid.

Glow tubes were later known as Geissler tubes after Heinrich Geissler, the German glassblower who greatly improved the glow tube by inventing a more effective form of vacuum pump in 1855 to remove the air inside. In the 1850s and later, Geissler tubes and lamps based on the principle of electrical discharge were continuously improved in many different forms. Variations included the mercury-arc lamp of 1901, which produced a very bright arc suitable for outdoor lighting, and neon and fluorescent lamps similar to the kind used today.

However, one branch of experimentation with glow tubes led to their use for purposes other than lighting. In 1855, J. M. Gaugain in France discovered that when a Geissler tube was fitted with two different-sized electrodes, an alternating current passing through it would be "rectified," that is, it would be converted into pulses of direct current flowing in one direction only. The effect of the tube was to act as a one-way valve, allowing the positive phase of an alternating current to pass, but blocking the negative phase. Again, the effect was merely a scientific curiosity at the time. A further elaboration of the Geissler tube was the invention of the cathode ray tube (CRT) by William Crookes around 1875. (Eugen Goldstein in 1876 had begun to call the flow through a vacuum "cathode rays," a term that stuck even after the electron was discovered.) Crookes's main discovery was that in a glow tube, many of the electrons missed the anode target, flying past it on all sides to strike the glass wall of the tube. Under certain conditions, the glass would fluoresce, or glow, and at that point the shadow of the anode could be seen. Others found that the straight-line path of the cathode rays could be warped or even reversed by a nearby electric or magnetic field, such as a magnet held close to the outside of the tube. Karl Ferdinand Braun in 1897 used these effects to make a sensitive measurement device, called an oscillograph, to be used for measuring varying electric currents. By using the electric field of the current to be measured as the modulator of the beam path, the device could display an image of an electric wave in the form of a visible "trace" on the screen. In 1895, Wilhelm Röntgen found that the Crookes tube also emitted some kind of unknown rays (he called them X-rays) that could, unlike the cathode rays, penetrate the glass walls of the tube. Subsequently, the Crookes tube began to be used in medicine, and along with gas-discharge lighting, this constitutes the second major application of electronics technology. Still, however, the word "electronics" was not yet in currency.

The immediate inspiration for the device that inspired the field of electronics appeared in 1880. Thomas Edison was developing a system of electric lighting using a technology that was quite similar in some ways to a glow tube. In his tube, or bulb as he called it, were two closely spaced electrodes connected by a short length of carbon, known as the filament. Electricity passing from one electrode through the filament and completing its circuit through the other electrode caused the filament to heat up and begin to glow. In a near-perfect vacuum, the filament material would not burn. Edison was free to use ordinary carbon, a material with a very high melting point. Such "incandescent" lighting tubes had been produced for many years, although Edison is usually given credit for commercializing the electric lighting system based on the principle. While his priority as inventor may be challenged, his commercial success cannot: the companies he founded or licensed became the world's largest electrical manufacturers

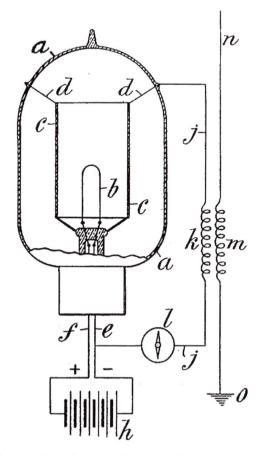

The Fleming "valve," or diode, from a 1905 patent. The device consists of a modified light bulb with a cylindrical anode (c), shown here in a cutaway section to reveal the interior. The positive half-cycles of radio waves captured by antenna (n) pass through the circuit from filament (b) to anode (c), causing a deflection in the needle of a galvanometer (l). This circuit acts as a sensitive detector of radio-telegraph signals. U.S. Patent 803684.

(General Electric, GEC in Britain, AEG in Germany, etc.) or electric power providers. In 1883, he was experimenting with a bulb that had a third electrode, not connected to the filament electrodes. He noticed that when this separate electrode was connected to the positive terminal of the circuit, some of the current flowing in the filament would flow to the free terminal through the vacuum. Connecting the terminal to the negative side of the battery produced no flow. He patented the device and thought it might be useful in making electrical measurements, but soon lost interest and moved on to other projects. After seeing a demonstration of the bulb, John A. Fleming in England studied its one-way "valve" action for alternating currents, passing a positive direct current or the positive phase of an alternating current from the filament to a "plate" electrode located nearby. He patented his version of the device in 1904 and proposed that it be used to rectify high-frequency alternating currents. When put in a circuit so that the negative terminal was attached to an antenna to capture radio-frequency electromagnetic pulses, and the output of the positive terminal linked to an indicator of some sort (such as a galvanometer), the valve could become the basis of a fairly sensitive radio receiver. Radio, called "wireless telegraphy" at this time, had been in existence only a few years. It used spark coils or high-frequency alternators to produce radio waves, while detecting them was accomplished by using a variety of electromechanical devices sensitive enough to respond to radio-frequency energy. The Fleming valve, as it became known, was a very sensitive, nonmechanical means to detect radio waves.

THE AUDION AND THE ADVENT OF "ELECTRONICS"

The elegance of the Fleming valve as a radio detector led American inventor Lee de Forest to add a third element to the Fleming valve two years later, to create an extremely important new type of device. Between the filament and the plate, he inserted a length of wire wound into a serpentine shape. Later versions had the wire formed into a gridiron-like pattern, so it became known as the "grid." By applying a small voltage to the grid, de Forest's tube precisely regulated the flow from filament to plate. De Forest's experimental results convinced him that he had invented a much more sensitive form of radio wave detector. De Forest also believed that the new tube could actually amplify the incoming waves, rather than simply detect them. Although he patented this application in 1907, it was apparently not until 1912 that he was actually able to demonstrate such an amplifier. Further, when he did so, he discovered that the amplifier circuit would occasionally overload, resulting in it spontaneously beginning to generate radio-frequency energy. This unanticipated function was seen at first as a nuisance, but later others realized that it could be controlled and put to work to replace the huge radio-frequency alternators then in use. It may not yet be clear to the reader why this device, called the Audion, was so important, but the reason is that amplification had become the holy grail not only in radio, but also in several other fields of electrical engineering.

In the interim between the invention of the glow tube and 1907, the telegraph, telephone, and radio had been invented, all without the benefit of electronics. Yet in the case of the telephone and radio, and to a lesser extent the telegraph, electricians eagerly sought a way to extend the maximum distance that they could transmit signals along a wire or through space. Telephone inventors called this sought-after invention a "relay," the term used in telegraphy to denote an electromechanical device that received an incoming, weak signal and used it to re-create a strong, clear outgoing signal. The relay was essentially an automatic telegraph key, which used the remaining energy of incoming electrical telegraph pulses to activate a sensitive electromagnet, which opened and closed a separate circuit to re-create the same pulses, but at a higher voltage and current. Telephone signals could not be relayed this way, because they consisted of complex waveforms rather than simple on-off pulses. Radio signals were originally

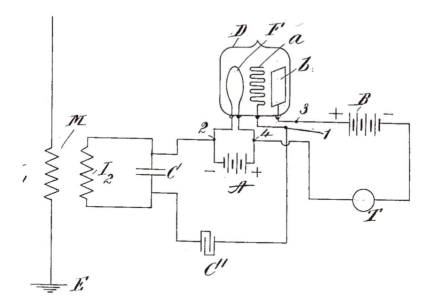

Lee de Forest's 1907 "Audion" was a modified form of the Fleming valve. Inside its evacuated glass envelope (D), current could pass from the filament (F) to the anode (b). The grid (a) acted as a control, impeding the flow to a greater or lesser degree depending on the voltage supplied to it. The output of the circuit can be heard on a telephone receiver (T). Later, de Forest and others developed more effective tubes and circuits to use the Audion as an amplifier. U.S. Patent 879532.

just high-frequency telegraph pulses, but by 1906 could also be in the form of voice signals (it was then called "wireless telephony"). Radio inventors sought ways to transmit both kinds of signals at ever-higher power levels, and also researched ways to detect extremely weak signals from distant stations. The sought-after telephone relay, high-power radio transmitter, and sensitive radio receiver would eventually all be found in the form of circuits based on the de Forest tube.

While de Forest himself went into the business of selling Audions for radio receivers, he was soon broke. In 1912 he asked to demonstrate the device to American Telephone and Telegraph Company, and AT&T immediately sought to purchase the legal rights to the Audion. Engineers at AT&T's subsidiary Western Electric began improving the Audion within a month, and the company soon became one of the world's leading makers of tubes. Yet de Forest's trade name would disappear; in later years the Audion would be known as the "triode," because of its three internal elements, and was distinguished from the two-element Fleming valve, which became known as the diode. The term triode, along with later variations such as pentode, fell by the wayside, although semiconductor rectifiers are still known as diodes.

THE EARLIEST USES OF SEMICONDUCTORS

While the vacuum tube is thought of as an outdated technology today (a view that this book challenges), some do not know that semiconductor devices are almost as old. Scientists of the eighteenth and nineteenth centuries categorized many materials according to their electrical properties, including their ability to conduct electricity. The elements germanium and silicon, along with numerous chemical compounds, were classified as semiconductors, with a level of resistance to the flow of electricity through them falling somewhere between good conductors and good insulators. These materials are still known as semiconductors today, although this terminology does not meaningfully describe the properties that make them useful in electrical circuits. The first of these useful properties that scientists discovered was the response to light of certain semiconductors. Willoughby Smith in 1873 observed that selenium's resistance dropped greatly when it was exposed to ordinary sunlight. In fact, a selenium "cell" or battery could be built that generated a small current when exposed to light. P. Nipkow used an array of such photocells in 1884 to demonstrate the principle of television in a very crude form (television imaging and display devices are discussed in greater detail later in this chapter).

Just a few years after radio appeared, semiconductor-based devices were first used as radio wave detectors. H. C. Dunwoody in 1906 discovered that a crystal of carborundum, the trade name of a silicon carbide abrasive invented in the 1890s, acted as a detector of radio waves. However, the crystal had to be mounted between two adjustable springs, through which the signal also passed. The exact reason why the springs were needed was not well understood, but experimenters discovered ways to adjust the springs to get a reliable radio detector action. Later, pure silicon or "galena" (lead sulfide) crystals became more popular as radio detectors. One end of a galena crystal could be permanently soldered into a brass cup and mounted on a plate. An adjustable bracket above the plate held a fine wire, the "cat's whisker," which had to be carefully positioned so that it touched the best spot on the crystal. That spot was found by trial and error. In all these crystal detectors, the interface or point of contact between the crystal structure and the wire or electrode was the key to operation.

A Simplified Explanation of Semiconductors

The usefulness of semiconductors in electronics stems from the structure of the atoms that make up semiconductor crystals. Carbon, silicon, and germanium, three common semiconductors, have four electrons in their outer orbitals (the top "shell" of orbiting electrons), so that when they are melted and refreeze, they form organized crystal structures or lattices. Mixing in ("doping") phosphorus or arsenic disturbs this structure, giving the crystal extra electrons that convert the crystal from an insulator to a conductor. Since electrons carry a negative charge, this type of crystal is known as an n-type. Doping the crystal with boron or gallium also turns the crystal into a conductor, but it does so by leaving it with a deficit of electrons, known as "holes," which make the crystal positive or p-type. Creating a junction between an n-type and a p-type crystal has the surprising result of creating a useful electronic device, called a diode, that conducts electricity in one direction because of the polarity of the crystals. Diodes can be used in electrical circuits to convert or "rectify" alternating current to direct current. They are also useful in radio receivers, where their rectifying action is known as "detection." In certain kinds of radio transmissions such as AM broadcasts, the received signals consist, in effect, of a varying direct current superimposed on a high-frequency alternating current. The diode blocks the alternating current portion and passes the rest along to the radio's amplifier.

Creating a sandwich of two back-to-back diodes creates a junction transistor, which can take the form of an N-P-N structure or a P-N-P structure. In this type of transistor, one outer layer is called the emitter, the center layer is the base, and the second outer layer is called the collector. Since the purpose of a transistor is to act as a switch or amplifier for current, the device is connected to a power supply such as a battery to provide the emitter with a source of current, while the collector serves as the output. If the base is left unconnected, no current can flow through the transistor. But if the base is also connected to the power supply, a little current will flow from the emitter through the base. Instantly the transistor will be "switched on," and a much larger flow will occur from the emitter to the collector. A more complex circuit can be added to set the amount of current flow through the base, and by doing this the emitter-collector flow can be regulated or modulated, allowing the transistor to employ a tiny flow to control a large flow. A transistor can then act as a simple on-off switch, as it does in computer logic circuits, or it can be used to amplify the signal from a telephone or microphone.

From the middle of the 1950s onward, it was known that junctions made from gallium arsenide emitted light (although it was only much later that usable lasers and light-emitting diodes, or LEDs, were made this way). Explaining this phenomenon introduces another set of terms. Free electrons moving through the semiconductor crystal have a fairly high level of energy and are said to be in the "conduction band." When an electron meets a hole in an atom, it falls into it and tends to stay there. The holes are in atoms in the lattice where an electron would normally be, and when a free electron falls in, it returns to a lower energy state. Its extra energy is released in the form of a photon of light. When the energy difference or "band gap" is small, as it is in silicon, the light is released at the invisible infrared frequencies. When the band gap is large, the emission is visible. In the type of diode used for switching or rectification, most of the light is absorbed by the diode itself. Light-emitting diodes are constructed so that most of the light radiates outward. The device is usually mounted in a small reflector cup to help direct the light, and the whole assembly is packaged in translucent plastic. A semiconductor laser uses much the same principle, using "heterostructures" or junctions of materials with widely varying band gaps, and employing mirrors or other means to reflect light emitted from the junctions in order to stimulate the laser effect.

PIEZOELECTRICS

Pierre and Jacques Curie in 1880 were apparently the first to measure the appearance of small voltages when pressure was applied to certain kinds of mineral crystals, such as so-called Rochelle salts. Thinking this was a "new" kind of electricity, this was dubbed the piezoelectric effect, a name that stuck. Somewhat later, it was also discovered that the reverse process also could occur: when such crystals were subject to an electric field, their lattice structures would be strained. This research ultimately led to two major classes of practical devices before 1945. The first included microphones, earphones, phonograph cartridges, and underwater sound transducers (for detecting submarines). The second was the use of vibrating quartz crystals as precision frequency regulators for electronic equipment, such as radar systems. In the postwar period, piezoelectric device research was revitalized by the discovery of a new class of ceramics with heightened piezoelectric properties. Interest in these materials corresponded with a growing interest in fields such as ultrasound, the use of extremely high-frequency sound waves for various purposes such as medical imaging and surgery, the cleaning of metals, and locating schools of fish in the sea. Much later, tiny piezoelectric devices became part of research in micromechanical systems (discussed in Chapter 6), with piezoelectrics performing duties such as the pumping of liquids on a miniature scale. By the early twenty-first century, piezoelectrics were commonly used as sensors to detect motion, impact, pressure, and strain. Many automobile car alarms, for example, used piezoelectric sensors to detect the distinctive vibrations emitted when automobile glass is broken. Piezoelectrics were also used as pickups for amplifying the sound of acoustic guitars. Another important application of the late twentieth century was the quartz watch, which used a piezoelectric device to generate an accurate electric signal that was then used to regulate the speed of the watch mechanism. But in the first decades of the twentieth century, this technology remained largely a curiosity.

THE EXPANDING USES OF ELECTRON DEVICES

Thus, by about 1915, radio-frequency rectifiers based on semiconductors and vacuum tubes, and amplifiers based on three-element vacuum tubes, were available and in use, as were various other types of tubes such as the CRT. Radio, which had been launched based on electromechanical technologies, was transformed into an electronic technology by about the end of World War I. Spark-gap wave generators and the massive, high-frequency alternators were dispensed with, and were replaced by more compact and usually less expensive vacuum tube transmitters. Telephone service was also greatly aided by the introduction of vacuum tube "repeaters" by 1915, when AT&T launched the first coast-to-coast long-distance service. The role of telephone companies in the history of electron devices continued in later years, and pausing here to explain this may make it easier to understand many later developments in electronics. Shortly after establishing long-distance service, one of the goals of telephone companies around the world became the automation of the process of connecting one telephone customer to another. Alexander Bell's original system used human operators working at central stations, who manually connected callers by physically plugging wires into a switchboard. The process was slow and expensive, especially for stations with more than a few hundred subscribers. Others proposed using a new type of telephone set to allow customers to connect the calls themselves. The new set used keys or a rotary dial to connect to the central station and select the desired connection. Automatic switching devices detected incoming information from the telephone (which consisted of DC pulses transmitted in a particular order) and used "relays" (a device physically similar to the telegraph relays described earlier in this chapter, but acting as an automatic switch rather than an amplifier) to route the call to the desired party. Between the 1930s and about 1960, ever-more-sophisticated automatic switching equipment, based on relays, gave customers the ability to connect local and long-distance calls, but this came at the expense of great complexity and rising maintenance costs. Switching equipment, almost never seen by the public, became the heart of telephone service, and its importance helps explain why telephone companies were so important in the history of device technology, because in later years they sought electronic replacements for conventional relays and switches.

THE "MISSION CREEP" OF VACUUM TUBES IN THE 1920S AND 1930S

In the two decades or so between the end of World War I and the beginning of World War II, the use of vacuum tube and semiconductor technologies expanded well beyond the boundaries of telephony and radio broadcasting. Beginning in the 1920s, for example, phonograph companies began recording their songs using new "electrical" disc recorders in the studio. In place of the old acoustic horn, these machines used a carefully designed electromagnetic cutting head, mounted on a lathe-like arm. The cutter was fed by microphones through a vacuum tube amplifier. The system was so appealing that by the mid-1920s most record companies adopted this technology. On the consumer side, the story was different. The Great Depression delayed the introduction of vacuum tubes in home phonograph systems, but by the late 1930s most phonographs were equipped with electromagnetic pickups, and shared an amplifier and loudspeaker with the radio receiver.

The late 1920s also saw vacuum tube technology applied to motion pictures to create the "talkies." The first talkies used large phonograph records, played in synch with the theater projector and amplified through powerful vacuum tube amplifiers and loudspeakers. General Electric, Western Electric, the Radio Corporation of America (RCA), and others introduced low-noise tubes and powerful amplifiers to suit this market. Just a few years later, these same companies and others including Pathé in France and Tri-Ergon in Germany began offering equipment to record the soundtrack directly on the motion picture film in the form of a visible record instead of a mechanical groove. Playback required the development of a suitable light-sensitive device to "read" the soundtrack. Several inventors proposed using the existing type of selenium photocell for the task, but these were generally not sensitive enough to give good results. A better solution was the "phototube," a light-sensitive vacuum tube (discussed later in this chapter).

Radio communication also began to expand to serve new needs. It had initially been provided mainly for ship-to-shore communication, but by the 1930s it was also commonly used on aircraft. This seemingly straightforward transfer actually required considerable improvements to make radio transmitters and receivers lighter, smaller, and more rugged. The military was becoming increasingly dependent on radio communication, and would eventually install it on virtually every ship, airplane, tank, submarine, and ground station. Radio "broadcasting" had also emerged in the 1920s, and by 1930 it was the basis of local, national, and international communications networks in virtually every country around the world. Particularly in the United States, where there were numerous stations in competition, supplying tubes and other devices for broadcasting became a major industry. The term "electronics" now came into regular use, and although it is not clear who first coined it, its popularity was reflected in the founding of the technical journal *Electronics* in 1930.

RADAR AND THE RISE OF THE MILITARY IN DEVICE RESEARCH

The military, particularly in Great Britain and the United States, took an early interest in the technology that became known as radar. Radar gets its name from the term "radio detection and ranging." Scientists had known since the late nineteenth century that radio waves could be reflected, but it was not until 1924 that Edward Appleton demonstrated a practical use for the phenomenon when he used a radio transmitter to measure the height of the ionosphere, measuring the time it took for the wave to be reflected back to the transmitter and calculating the distance based on the speed at which radio waves traveled. In 1934 or 1935, a French firm, Compagnie Générale Transatlantique, used this method with a high-frequency transmitter to detect ships hidden by fog or darkness. The tube they chose, eventually called the Magnetron, had its origins in studies of the paths of electron motion undertaken by J. J. Thompson in the 1890s. Thompson explored the paths that electrons took as they flew under the influence of a strong magnetic field. Later, the Swiss physicist Heinrich Greinacher invented a tube with a central cathode, surrounded by a curved or tubular anode. When a strong magnetic field was applied outside the tube envelope, the electrons began to circulate inside the tube rather than striking the anode. This circulation would cause very high frequency energy to be generated inside the tube. Although the German physicist Erich Habann demonstrated a high-frequency radio transmitter based on such a tube in 1921, Albert Hull of General Electric in 1920 investigated the principle and designed a tube suitable for use as an AM transmitter. It was Hull's tube, called the Magnetron, which was remembered in later years as the genesis of an important line of microwave generator tubes. GE did little with the Magnetron, but the basic design was continually improved in the 1920s and 1930s. Engineers in Germany and Great Britain discovered that with stronger magnetic fields and higher operating voltages, the tube exhibited some interesting properties. Although it was poorly understood just how it worked, the tube could provide high levels of power output compared to conventional "gridded" tubes. The reason had to do with the interaction between the circulating electrons and the external magnetic field. The interaction of the fields caused some electrons inside the tube to slow down and others to speed up. As a result, they would become bunched. The bunches would then transfer their energy to the output circuit of the tube in the form of powerful microwaves.

An important elaboration of the Magnetron was the Klystron, invented in the late 1930s by two brothers, Sigurd and Russell Varian. In their tube, an electron beam passed straight through a long, empty cavity. Along the way it interacted with modulating fields, resonating inside doughnut-shaped chambers. The electrons formed bunches, and then transferred most of their energy by electromagnetic induction to a tap near the end of the tube. The remaining energy in the electrons was wasted in the form of heat at the tip of the tube, which had to be cooled to keep the tube from overheating. The tube was a powerful microwave generator, and its operation was easier for engineers at the time to understand than that of the Magnetron.

THE CAVITY MAGNETRON AND RADAR IN WORLD WAR II

Until the invention of the cavity Magnetron and the Klystron, researchers were limited to systems that could produce radar pulses of several kilowatts or less. In 1939 John T. Randall and Henry A. Boot at the University of Birmingham in England produced, apparently by accident, an improved Magnetron that became the basis of an important line of research in the radar field. The tube, called the cavity Magnetron, showed promise of greatly exceeding the power output of conventional tubes. The name was derived from the shape of the anode. Instead of using one or more curved plates as others had done, the anode was made from a short cylinder of solid metal, into which a series of holes had been drilled. In the center was a large central hole, into which the cathode in the form of a metal wire was inserted. Then passages were cut from the center hole to each of the surrounding cavities. When current was flowing in the tube, the electrons interacted with the anode block, and strong resonating fields were set up, the frequency of which depended on the size of the cavities. Electrons moving toward the anode would also interact with an external magnet, which prevented them from reaching the anode but caused them to circulate as a sort of spinning cloud in the central cavity. The magnetic field emanated by this cloud interacted with the anode fields, setting up an oscillating magnetic field at each cavity. These fields further interacted with the electron cloud, drawing energy from it and becoming increasingly stronger. The size and spacing of the cavities determined the frequency of the fields created there, and the tube could easily be constructed to resonate at microwave frequencies. A simple loop of wire inserted into the side of one of the cavities tapped off some of the microwave energy, which was then carried off to the radar transmitter. After working on a cavity Magnetron for 10-cm wavelengths for some time, the British researchers turned over their design secrets to engineers in the United States, so that they could concentrate on

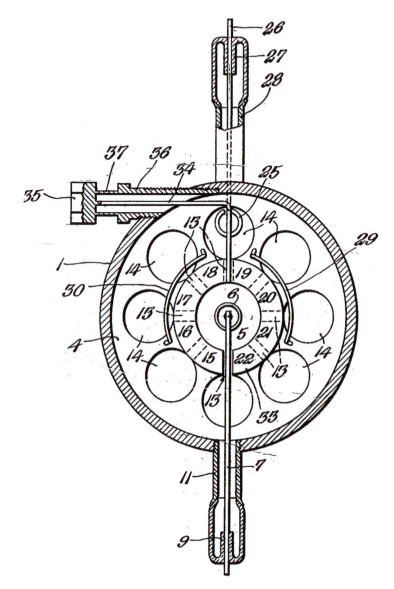

Percy Spencer's high-efficiency Magnetron of 1941 used a strong magnetic field to send the free electrons in a vacuum tube into spinning, circular paths. Shown here in a cutaway section, looking down, the size and spacing of the multiple resonant cavities (14) cut into the anode block (4) contribute to the effect. The effect is to produce strong waves of high-frequency energy, which is tapped off by a coupling loop (25). U.S. Patent 2408235.

more immediate needs. Bell Telephone Laboratories and others improved the British Magnetron, delivering for example a tube for 20- to 30-cm wavelengths a few years later that was capable of delivering a pulse power of 750 kW. This kind of power had never been available at such short wavelengths before, and the resulting radar systems proved extremely important in the Allied victory.

Meanwhile, work continued on the Klystron, which was eventually considered part of a new category of devices known as traveling-wave tubes (TWTs). Bell Telephone Laboratories contributed to early work on new types of TWTs, resulting in a practical model made by Rudolf Kompfner, A. W. Haeff, and John R. Pierce of Bell Telephone Laboratories in 1943. The tube, which was originally aimed at high-power microwave generation, turned out also to be an amplifier of extremely wide bandwidth, opening up nonradar possibilities for it.

THE MILITARY AND MINIATURIZATION

Radar tubes were also at the heart of one of the first efforts to produce miniaturized electronic systems. W. S. Butement in Britain proposed something called the proximity fuse in 1941. Inside an artillery shell, a tiny radar set transmitted pulses and measured the delay between outgoing waves and incoming reflections. As long as the shell passed somewhere near its target, the delay would decrease to a certain predetermined value. When that happened, the circuit triggered an explosion. The engineers who proposed this system realized that it would require tubes much smaller than those being used in conventional radar, and the tubes would have to be able to withstand the enormous stress of being fired from a cannon. The British, who started this research, later realized that they did not have the resources to finish the development of the fuse and put it into large-scale production, so they asked the Americans to do it for them. The shells were ready for service by 1941, and were later used to combat the extremely fast, unmanned German "buzz bombs" then being launched against London. The kill rate for buzz bombs soon rose to nearly 100 percent. Over 150 million proximity fuses were manufactured by the end of the war by U.S. companies including the Raytheon Corporation. While the proximity fuse was one of the most impressive technical accomplishments of the war, it also signaled the beginning of an intense effort on the part of the military to miniaturize the electronic devices used in aircraft and missiles. This focus on miniaturization would, in later years, contribute directly to the success of technologies ranging from the transistor to the integrated circuit.

THE BEGINNINGS OF DISPLAY
AND IMAGING DEVICES

Numerous electrical and electronic systems require some kind of device to act as an information display or indicator. Early nineteenth-century proposals for the telegraph relied on electromagnetic, alphanumeric indicators using one or more moving needles to point to a letter or number, and needle indicators were common through the twentieth century for various kinds of electrical meters and gauges. Some of the most familiar included resistance-type needle indicators used as fuel-level indicators in automobiles and the electromagnetic volume unit indicator, or "VU" meter, used on audio equipment. The incandescent lamp, when it became commonly available in the late nineteenth century, also became a common form of indicator. Single lamps could display "power on" or other simple information, while arrays of lamps could be used to spell out words or create images. Most proposed television systems from the 1890s to the 1920s relied on arrays of lamps, along with ancillary mechanical devices, to form images. Small, gas-filled lamps of the type in use since the mid-nineteenth century were widely used in telephone switching stations to indicate the "busy" status of lines. In consumer electronics, one important type of indicator was the "magic eye" tube, developed in the 1930s. A form of cathode ray device, the magic eye illuminated phosphors on a circular target anode, located so that it could be easily seen outside the tube envelope. The size of the illuminated spot could be made larger or smaller to indicate volume or radio signal strength. Until the advent of inexpensive VU meters around 1960 (a technology that was itself replaced later by electronic displays), the magic eye was the standard form of level indicator for recorders and the more expensive radios.

CATHODE RAY TUBE DEVELOPMENT THROUGH
THE 1930S AND 1940S

The most important of the vacuum tube display technologies was the cathode ray tube. The CRT as demonstrated by Braun remained little used until the 1920s, when Hendrik J. van der Bijl and John B. Johnson at Western Electric developed a small CRT for use in electronic oscilloscopes. The tube, called the type 224-A, was probably the first in regular production. The envelope of this tube was flask shaped, with a relatively large, flat, round screen for easy viewing and measurement of waveforms. Through the end of the vacuum tube era, most CRTs for oscilloscope work used electrostatic deflection of the beam by means of two sets of plates inside the envelope. This provided a faster reaction time than an external electromagnet, but electromagnets gave better results for television, radar, and computer monitor screens that came later.

Oscilloscopes constituted a small but steady market for CRTs through the 1930s, as the electronic oscilloscope became a standard feature of many industrial and university electronics laboratories. When RCA decided in the early 1930s to devote its considerable resources to television, engineers chose the CRT for its display. CRT development, thus well underway by the time World War II radar research began, benefited television equally. During World War II, the U.S. government alone spent a remarkable $2.7 billion on radar systems for use by the military, some of which resulted in larger, brighter, and more responsive CRT display screens. At first there were few commercially available CRTs, so the standard form of radar indicator was an off-the-shelf cathode ray oscilloscope. Early radars often displayed radar "reflections," representing objects detected by the radar as an unfilled circle on the screen with spikes (indicating radar reflections) projecting outward from it. A second variation was to show the spikes projecting from the bottom of the screen. The latter was known as the "A-Scope" style of presentation, but was superceded beginning in 1941 by the plan position indicator (PPI) radar display. An airborne PPI radar aimed at the ground, for example, would display the terrain and objects on or above it as if they were being lit by a huge lamp. This type of radar is credited to a joint effort between the U.S. Naval Research Laboratory and researchers in Great Britain. PPI radar still used an ordinary CRT, but in radar the image was refreshed less frequently than it was for television, so there was a desire for a CRT with phosphors that could continue to glow for up to several seconds. As a result, PPI radar CRTs used a technology known as P7 screens, first developed in Great Britain in 1938, which provided better image retention than standard oscilloscope CRTs of the time.

CRTS FOR TV IN THE 1940S

Most television systems proposed before the early 1930s constructed an image using a rapidly rotating disk with a spiral of perforations. A row of small light bulbs, flashing at the appropriate moments behind the disc, would form a complete image due to the eye's persistence of vision. However, the image quality was always poor. Electronic television using the CRT as a display seemed like a better alternative, and such a system was demonstrated in Germany in 1935 and used for several years before World War II.

Various inventors and companies in the United States and Europe were moving toward electronic television, but none was as successful as the Radio Corporation of America, which began broadcasting in 1939. Commercially available RCA CRTs in the 1930s were 5 or 9 inches in size and used green (or, later, yellow) phosphors. A 12-inch screen was introduced by 1939, at which time the Philco and National Union companies were marketing CRTs in the United States, while Baird in England was already making 12-, 15-, and 22-inch "Cathovisors." Dumont and Cossor (an English firm) were also marketing electrostatically deflected tubes for television. The former had 14- and 20-inch CRTs in production by 1939.

While commercial television was put on hold during the war, some TV-related research continued. RCA, for example, developed a remote-control flying bomb guided by an onboard television camera, which transmitted signals back to a control station located nearby. The flying bombs saw limited use in the Pacific theater.

IMAGING

For the purposes of this discussion, an imaging technology is any device used to detect visual information, but this can range from simple light detectors to more complex arrays of thousands or millions of individual devices used in today's electronic cameras. Individual selenium cells were used experimentally in the late nineteenth and early twentieth centuries as light sensors for purposes such as automatic door openers. Because of its relative insensitivity, it was replaced in many applications by phototubes beginning in the 1930s. These tubes incorporated a light-sensitive photoemitter, which was usually the semiconductor material silver-oxide-cesium coated onto a metal cathode plate. The cathode would emit a weak stream of electrons when struck by light, which would be collected by the anode. A triode amplifier was usually included in the same tube envelope to raise the output to levels suitable for subsequent amplifier stages. Similar phototubes were enthusiastically promoted for use in industrial applications in the 1930s and 1940s. Often called "electric eyes," the tubes could detect the presence or absence of objects based on light reflected from them or shone through them, and in this way electronic circuits could perform rudimentary quality control or other simple industrial applications.

CAMERA TUBES

The gathering of more complex images in electronic rather than photographic form was an essential part of the development of television. RCA hired Vladimir Zworykin, a Russian immigrant who had patented a television system using a Braun tube (CRT) screen and a special electronic camera tube called the Iconoscope in 1923. The camera tube utilized a target plate made from a sheet of mica insulator, onto which tiny droplets of photoemissive material had been deposited. Each droplet corresponded to a single pixel of video information. Zworykin revisited this design beginning in 1929 at RCA. He tested the new tube, beginning in 1934, utilizing the now-familiar mosaic or array of photo-emissive picture elements (pixels), consisting of a silver-oxide base with a cesium oxide coating. Each pixel was scanned by an electron beam. Light striking a granule of photoemissive material caused it to emit electrons and become positively charged. Thus the array initially captured an image on its grid of pixels, each made more or less positive according to the intensity of light that had struck there. Behind the mica sheet supporting the photoemitter array was a metal "signal plate," and each granule formed a capacitor between itself, the mica insulator, and the signal plate. Then, when scanned by the beam, the globules collected electrons and their charge state suddenly changed. This caused a flow of electrons at the signal plate because of the capacitor action. As the beam scanned across the array, each pixel would cause a discharge into the signal plate, and the result was a television signal. The tube worked in lower ambient light levels than earlier television scanning methods.

The Iconoscope was the first camera tube that worked adequately with live television, although it still required very high levels of stage lighting to have adequate image contrast. It was put into service in 1939 in RCA's first TV broadcasts. In 1939, Albert Rose and Harley Iams of RCA co-invented a new type of television camera tube called the Orthicon. Unfortunately, the war had already ended television broadcasts, but RCA continued to develop the tube under military funding. It was eventually used in various military equipment, such as a camera-guided, remote-control missile that saw limited use late in the war. The Orthicon contained a sensitive, photoemissive array, onto which the desired image was focused. Photons striking the array caused it to release electrons according to the intensity of the light, and these electrons were collected at the rear of the tube by a collector plate. The scanning beam recharged the granules in succession, as in the Iconscope, and when that occurred each pixel sent a signal pulse to the common signal plate terminal. Excess electrons from the beam were reflected to the collector plate. The tube used a higher voltage difference between the image mosaic and the collector plate for the emitted electrons. The result was that more of those electrons streamed back into the interior of the tube to strike the collector. This reduced the incidence of them returning to the mosaic, resulting in image "noise." RCA achieved additional reduction of this secondary emission by keeping the mosaic at the same voltage potential as the electron gun, so that when the electrons struck the photoemissive granules they did so more gently. While it was more sensitive than the Iconscope, the low-velocity electron beam was hard to focus, so work continued through the 1940s to improve it.

The Orthicon would return after the end of the war to replace the Iconoscope for television cameras. Rose, Paul Weimer, and Harold Law of RCA improved it, reintroducing it as the Image Orthicon. The Orthicon took advantage of secondary electron emission rather than avoiding it. The photoemissive mosaic emitted electrons toward the interior of the tube, where they were accelerated by an electric field and struck a target plate a short distance away. When this occurred, the electrons were displaced from the target, and a "charge image" was created, consisting of an array of positively charged regions. This charge image was several times stronger than the original image because of the acceleration. Then the scanning beam passed over this target, returning each region to a neutral charge state and reflecting any excess electrons backward, to be collected at the rear of the tube. Rather than disposing of this reflected beam, however, the Image Orthicon derived the signal from it, since the beam varied in exact synchronization to the state of charge of each region on the target. The result was a tube with even greater sensitivity to light. The Image Orthicon quickly became the standard TV camera tube and remained so for two decades. Although it was later replaced by semiconductor devices, the tube is memorialized in the name of the Emmy award (originally known as the Immy) of the National Academy of Motion Picture Arts and Sciences.

COMPUTERS AND ELECTRON DEVICES

Today the field of computing is so heavily dependent on electronic devices that it is almost impossible to discuss the two separately. However, like the phonograph, telephone, and radio, early computers were entirely free of electronic devices, and in many cases free of electrical circuits as well. Modern computers had their origins in mechanical devices used to make simple calculations, in specialized machines created to solve complex differential equations, and in equipment used to collect, sort, and store numerical information. A number of European mathematicians invented machines to calculate numbers beginning in the 1500s. By the mid-nineteenth century, in the midst of the Industrial Revolution, mechanical calculators small enough to fit on a desktop were in use in all sorts of businesses and government offices for accounting purposes.

In the 1920s, Massachusetts Institute of Technology physicist Vannevar Bush proposed that complex equations involving multiple variables could be simulated using so-called analog computers. An equation to be solved was, in a sense, simulated by translating its variables into gears, cams, or other mechanical components. Analog computers, because they handled very complex equations easily, played a vital role in tasks such as flight simulators (first used in World War II) and artillery aiming for years to come. They would also be translated into electronic designs, although these apparently played only a small part in reshaping the course of device technology. More important in this respect were digital computers.

Today's digital computers incorporate some ideas from analog predecessors, combined with a few others. An important influence on computer design was the Hollerith tabulating machine, introduced in the 1890s. Herman Hollerith proposed storing population census information on punched cards. His machine could then sort the cards and extract lists of people with certain characteristics. While some of Hollerith's machines used electric motors, their sorting functions still relied on purely mechanical processes. Digital computers also drew upon the earlier generation of business calculators used for simple arithmetic. In the 1920s it was common for researchers working on very complex mathematical problems to divide the problems up into discrete arithmetic tasks and distribute the tasks to human calculator operators (called "computers"), who performed the arithmetic and tabulated the results. This was all done by hand or mechanically until 1937, when Bell Laboratories engineer George Stibitz constructed a calculator that used electrical telephone relays to do the calculations. The relays were arranged into circuits so that an operator could, say, press buttons instructing the machine to add two numbers and see the results displayed as a glowing light bulb. Bell Labs management encouraged Stibitz to construct a more complex model capable of receiving and transmitting data across telephone lines to a teletype machine. A novelty at the time, the relay calculator operated so much faster than a mechanical calculator that researchers took notice.

By the late 1930s, many of the other features that would be used to build the first modern computers were already in place, such as binary mathematics. There was still little call for these devices. As World War II neared, that began to change. A professor at the University of Iowa, John Atanasoff, decided to construct an improved binary calculator in 1939. He proposed feeding data to the calculator with punched cards, with the actual arithmetic taking place in electrical circuits, and intermediate results being stored optically on cards written by the machine. Numbers in "memory" would be stored as charges on a bank of capacitors, mounted on a spinning cylinder. Others, such as Konrad Zuse in Germany, were working along similar lines.

A more complex computer, built during World War II by IBM and Harvard University, was known as the Harvard Mark I. A program was fed to the computer on punched paper tape, and data was input via punched cards. Actual calculations were performed in the computer's vast assemblage of relays, clutches, and rotating shafts. Meanwhile, the British constructed a programmable computer called Colossus to help them crack German secret codes. Instead of relays, this computer used vacuum tubes, wired into circuits so that they performed as relays (that is, fast-acting switches), but had no moving parts. Between 1943 and 1946, another computer called the Electronic Numerical Integrator and Computer (ENIAC) was built at the University of Pennsylvania. By this time, vacuum tubes were beginning to displace relays because computer designers recognized that the switching speed of a tube was much faster than that of a relay. As computers grew more complex and the number of relays or tubes grew, switching speed became even more important. The drawback to tubes was that they were less reliable than relays and had to be regularly replaced. By one estimate, fifty of the ENIAC's approximately 18,000 vacuum tubes had to be replaced every day.

The ENIAC was reprogrammable only by physically rearranging its circuits, though within a year or two its creators proposed developing a "stored program" computer to make reprogramming more practical. This, however, required some kind of memory system to store instructions and sometimes intermediate data temporarily. Most subsequent computers also adopted the stored program approach, and used vacuum tubes for calculations and mercury delay lines for the temporary storage of instructions. The mercury delay line was essentially a glass tube filled with mercury, with quartz crystals at each end. Electric pulses sent to one end caused the quartz to vibrate, and the vibration was transmitted through the mercury and picked up by the crystal at the opposite end, which generated a new output pulse. The transfer through the mercury took a moment, so the effect was to delay the pulse. The delay was long enough for the device to be useful as memory. However, the data had to be continually looped through the line in order to keep them in memory, and when the power was switched off, the data were gone.

A vast number of individual devices was needed to satisfy designers of computer memory after about 1950. Vacuum tubes proved uneconomical almost immediately, and for many years, while computers remained short on memory, computer programmers were trained to write programs that used as little memory as possible. Where data did not have to be accessed at maximum speed, storage was accomplished using magnetic tapes, drums, and discs (the latter related to the hard discs in use today). Other devices briefly emerged as possible contenders for faster, random-access memory, such as the Williams tube. This was a modified form of the CRT that stored each bit of binary information as a bright or dark spot on the face of the CRT. The state of these spots could be detected by electrodes placed on the outside of the tube, because the charged phosphor, glass envelope, and electrode formed a capacitor. Because the glow of a phosphor persisted for a short time after it had been lit by the scanning beam, the tube retained information on its screen. RCA in the late 1940s also developed a vacuum tube–based memory device, the Selectron, capable of storing 256 bits of information. In simplified form, the Selectron consisted of a central cathode surrounded by up to 256 grid structures. Data would be written, erased, and stored by electronically selecting any of the 256 circuits and sensing or rewriting its state, on or off. Neither of these vacuum tube solutions proved economical, and so from the 1950s through the early 1970s most computers relied on what was called core memory. Harvard University's An Wang was the first to propose storing computer data using ferrite (iron) rings or cores, placed into arrays addressable by a matrix of fine wires. When subject to a positive or negative pulse from one set of wires, the core would retain a magnetic charge oriented toward one direction or another depending on the polarity of the pulse. Other wires could be used to "read" the state of the cores as necessary. Core memory, because it was assembled by hand, was expensive, and the relatively slow response time of the ferrite material made this type of memory slower than transistors. Core memory remained the dominant form of computer memory until its eventual replacement by the modern type of semiconductor chip.

THE INVENTION OF THE TRANSISTOR

Such was the history of electron devices in the nineteenth and early twentieth centuries. The rest of this work focuses on the period after 1947. That was the year that a device known as the transistor was invented at Bell Laboratories in New Jersey. It utterly revolutionized the field of electron devices, displaced the vacuum tube in most of its applications, and ushered in the computer age. The transistor is a semiconductor device, and in its early form it was similar in construction to the cat's whisker semiconductor diodes in use in radio. Bell Telephone Laboratories engineers in the late 1920s began investigating the properties of various semiconductors in the hope of eventually finding replacements for diodes and triodes. Some of the early work concentrated on copper-oxide rectifiers. Copper oxide, a semiconductor compound, was then coming into use to make diodes for rectifying currents, and could for example be used to provide DC power for telephone equipment. RCA had offered a copper-oxide "Rectox" diode in the early 1930s, which consisted of a stack of copper discs coated with copper oxide on one side, with lead discs in between. In Germany at the Siemens-Schuckert company, Walter Schottky had studied similar rectifiers, concluding that the rectifying action must take place at the junction between the copper and the copper oxide.

Most early proposals for a semiconductor replacement for the triode were based on the analogy between the existing form of the diode and the triode, the two principle types of vacuum tube that differed only slightly in physical construction. Researchers believed that a grid or a similar type of electrical control device could be employed to regulate the flow of electrons in semiconductor diodes or crystals. In 1938, for example, R. W. Pohl in Germany coauthored a paper describing how, in theory, one could make a replacement for the triode based on a solid crystal. He demonstrated the concept by using a crystal of potassium bromide. Electrons would be emitted from a point-contact wire, then moved toward an anode at the far end. A wire inserted into the crystal near the cathode was supposed to modulate the flow the same way that the grid modulated electron movement in a triode. The device actually worked, but the modulating action was very slow and it worked only when the crystal was heated.

Somewhat different in approach were the "field effect" devices proposed in 1926 by Julius Lilienfeld of Brooklyn, New York, and later by 1935 by Oskar Heil of Berlin. Both inventors used a sample of a semiconducting compound, sandwiched between two metal plates. One plate was connected to a source of current, and the other side was connected to the device output. Above and external to the semiconductor sandwich, and insulated from it, was a second plate or terminal to which a control or signal was fed. A voltage applied to the top plate created an electrostatic field, which was supposed to affect the semiconductor material, causing its resistance to drop. By varying the control field, the device would in theory perform like a vacuum tube triode. Whether these devices worked or not is unclear, and in any event these inventions faded into obscurity.

THE RETURN OF THE CAT'S WHISKER

Semiconductor versions of the triode, so-called field effect devices, and ordinary semiconductor diodes or rectifiers were thus all evolving simultaneously by the 1930s, a fact that complicates the history of the transistor's invention. At Bell Labs, work on semiconductors (some of it purely theoretical) was also proceeding, and it was focused on improving the metallurgical techniques for creating the semiconductor materials as well as investigating the properties of existing types of cat's whisker diodes. While the semiconductor cat's whisker diode had fallen into disuse in radio broadcast receivers during the 1920s, it was revived in the late 1930s for use as a detector in radar. The triode tube, which had served well as a detector of ordinary radio waves, performed poorly at the higher frequencies used in radar. The semiconductor cat's whisker detector, however, worked better in this applications, and it became the focus of considerable research at Bell Laboratories and many other places from the period just before World War II into the late 1940s. British and U.S. manufacturers made thousands of semiconductor diodes for use in military radars during the war. By 1942, General Electric had also gotten into this field, producing cat's whisker diodes made from crystals of germanium.

Although cat's whisker or "point-contact" radar diodes were massproduced during the war, they were still somewhat unreliable owing to the extremely critical placement of the point of the terminal wire on the surface of the semiconductor crystal. Slight changes in temperature or a tap on the side of the rectifier was likely to render it useless. The interface between the point and the semiconductor was known to be the key to its operation, but physicists could not adequately explain why. Perhaps for this reason, Walter Brattain, a researcher at Bell Labs, was more interested in the copper-oxide rectifier, where a large copper surface was mated firmly to the copper-oxide semiconductor material. After joining Bell Labs in 1929, he worked on these devices during the 1930s. Others at Bell Labs, such as Russell Ohl, thought that the key was in eliminating impurities from the semiconductor material. Their work, combined with that of several other researchers, would begin to bear fruit in a few years.

Even with its more robust mechanical design, the most problematic part of the copper-oxide rectifier was the interface between the copper and the oxide. Brattain reasoned that if a gridlike structure or electrode could be inserted into the junction, the diode might become a triode and be useful as a switch or amplifier. But he was never able to construct a functional device based on this idea. William Shockley, another Bell Labs researcher, tried much the same thing later, during 1939 and 1940, and even had Brattain

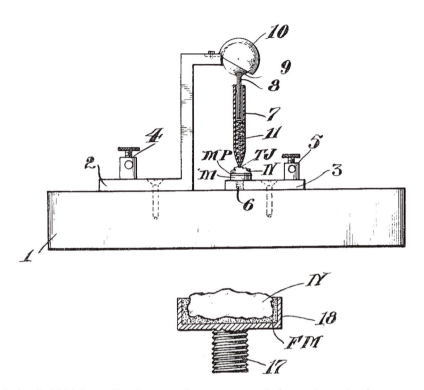

Pickard's 1906 "crystal" radio wave detector consisted of a small sample of semiconductor material (n) held between two metal contacts, a cup (m) and a wire "cat's whisker" (11). U.S. Patent 836531.

construct several models for him, but again the device failed to work. Shockley also experimented with a crude, semiconductor field effect device, but it too failed to work.

Then, around 1940 this research at Bell Labs took a new tack. Russell Ohl, who in 1939 had worked on equipment to produce ultrapure silicon ingots, was testing the resistance of a rod-shaped slice taken from one of his ingots when he found that it exhibited a much greater photoelectric effect than was possible with other photocells. There was something about this rod that made it special, and upon further testing, the rod proved to function as a rectifier, too. As it turned out, the larger ingot from which the rod had been cut had a region with a high level of impurities and a region with a lower level of impurities. The junction between the two formed the same kind of interface or barrier as the copper/copper-oxide junction in the copper-oxide diode. Because of the one-way "valve" action of the junction, a voltage applied to the rod (or the electrons dislodged by incoming photons) tended to move in one direction along the rod, providing this diode behavior. Further, the low-impurity and high-impurity regions had different levels of resistance and differing responses to light. Ohl and Jack Scaff, who participated in this research, named the two regions positive and negative, or p-type and n-type. The interface was called a p-n junction. This terminology is still used today to describe semiconductor devices. The two then worked to identify the impurities that resulted in these two regions, identifying aluminum and boron as giving rise to p-type, with phosphorus impurities resulting in n-type.

THE TRANSISTOR ARRIVES

Work on the field effect and semiconductor triode devices as well as semiconductor junctions was interrupted by the outbreak of World War II, but in the years immediately following the war, Bell Labs decided to emphasize research in solid-state physics, microwave physics, and electronics, including research into silicon and germanium semiconductor devices. William Shockley was to be the head of this effort, and in 1945 Brattain and another researcher, John Bardeen, were assigned the goal of discovering why the earlier field effect experiments had failed. Shockley and Bardeen both concluded that in theory the field effect device should work, and that something happening at the surface of the semiconductor material was

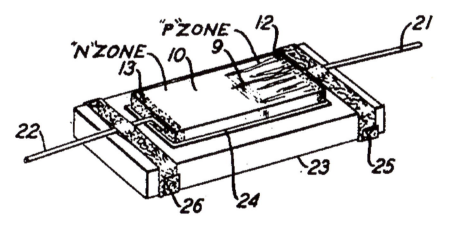

Semiconductor junction diode patented by Russell Ohl of Bell Laboratories in 1941. This device is a slice from a single crystal containing both n- and p-doped regions. Light striking the device generated a small current, making it useful as part of a "solar battery." U.S. Patent 2443542.

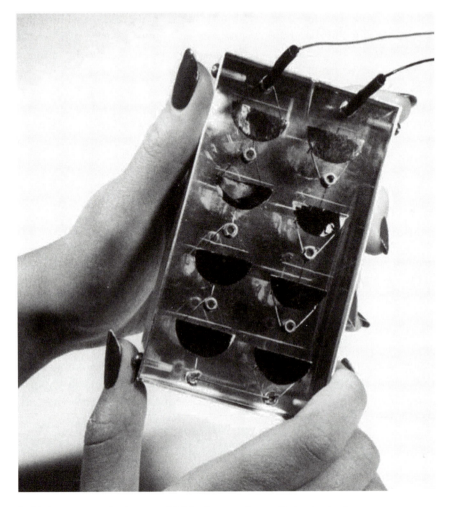

Bell Laboratories solar battery, 1954. Courtesy Lucent Technologies Inc.

preventing it from working. Research in the laboratory turned to the examination of the semiconductor surface. By 1946, Walter Brattain was working on the problem in the laboratory, and found that the field effect device would work if it were submerged in water. The electrolytic action of the water seemed to change the problems at the surface that prevented the thing from working. Bardeen made valuable suggestions, and the two seemed close to a semiconductor amplifier design using point-contact probes that contacted the semiconductor surface through a tiny drop of water. Following up on this experiment, Brattain was testing a block of germanium that he thought had a thin coating of oxide on its surface to insulate it. However, he had accidentally washed the coating off. When he jabbed probes into its surface to test its electrical resistance, he discovered that when touched by one positive probe and one negative probe, the block would amplify a voltage without the need for the water droplets. Bardeen and Brattain designed an improved model with the probes held in place very close together, which was demonstrated for the first time on December 16, 1947. The device relied on neither the field effect nor the semiconductor junctions developed earlier, but in most ways was simply a modified form of the cat's whisker. Nonetheless, it worked quite well. The two, along with William Shockley, posed for what became a famous photograph with the new device (see photo in Chapter 2 and "Nick Holonyak: On the Transistor Inventors").

Nick Holonyak: On the Transistor Inventors

Nick Holonyak worked at Bell Telephone Laboratories from 1954 to 1955 and later at GE, contributing to early semiconductor technology, LEDs, and other areas.

John [Bardeen] walked in and he asked me if I had seen *Electronics* magazine, and I said, "No." One of the grad students was standing nearby, and I sent him across the street to our main building. The young man came back with the April 17, 1980, *Electronics* magazine, and John and I were thumbing through it. And then he hit the famous picture—the one of Bardeen, Brattain, and Shockley. He was only as far away as you are, and he said to me, "Boy, Walter really hates this picture." I said to him at the time, "Why? Isn't it flattering?" . . . That's when he said to me, "No. That's Walter's apparatus, that's our experiment, and there's Bill sitting there, and Bill didn't have anything to do with it." . . . What John meant right then was very simple: Shockley didn't have a damn thing to do with that experiment and with the basic discovery of the bipolar transistor. . . . If you want to tell the truth, and you want history—From Bell's point of view, it's not necessary to say this because three people at Bell Labs invented the transistor. Bell Labs is this Mecca where these great things happen. Bullshit! It doesn't happen that way. It doesn't happen that way at all. These are all people. They're in constant pulling and tugging, and sometimes it's smooth, sometimes it's not so smooth. And history ought to know this because then young people really are seeing that this stuff is done by people, and it isn't done in some clean, antiseptic, beautiful way. It's done with guesses, with arguments, with all the foibles and peculiarities of people.

Source: Nick Holonyak, an oral history conducted June 22, 1993, by Frederik Nebeker, IEEE History Center, Rutgers University, New Brunswick, New Jersey.

Brattain and H. R. Moore soon used this transistor to build an audio amplifier, and by Christmas Eve of that year Bell Labs realized that it had achieved an important breakthrough. Now all it needed was a name. A committee met and suggested several names, including "semiconductor triode," "Iotatron," "solid triode," and others. Then they circulated a memo putting all their suggestions to a vote. The winning name was suggested by John R. Pierce, who combined the words "transfer" and "varistor" (the name of an earlier Bell Labs innovation) to make "transistor."

The transistor's discovery was announced in June 1948. Bell Labs promised that the transistor would soon replace vacuum tubes due to its simplicity and small size, but the device initially provoked little excitement in the electronics industry as a whole. Only the editors of *Electronics* magazine seemed impressed and put the transistor on the front cover of their September issue of that year, stating that "because of its unique properties,

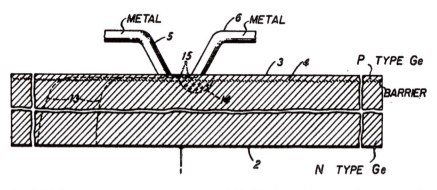

The Bell Laboratories point-contact transistor of 1947, depicted in a much more orderly way than the actual device. Two wires touched the surface of a germanium crystal with a thin p-type coating on an n-type substrate. When connected in a circuit a certain way, a small audio signal between one terminal (5) and the substrate (2) allowed a larger current to flow between the substrate and the second terminal (6). U.S. Patent 2524035.

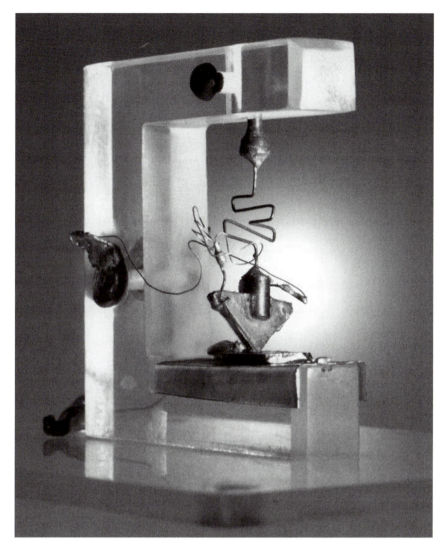

Laboratory model of the first Bell Laboratories point-contact transistor. Courtesy Lucent Technologies Inc.

the Transistor is destined to have far-reaching effects on the technology of electronics and will undoubtedly replace conventional electron tubes in a wide range of applications." Others were considerably less enthusiastic: the *New York Times* devoted a mere 4½ column inches in its radio news column to the invention. This limited enthusiasm is understandable given the fact that Bell Labs did little to explain why a replacement for the vacuum tube was important. Point–contact transistors offered a number of advantages over tube technology, including small size and efficiency, but they were also difficult to manufacture and could handle less power than tubes. As a result, in the years immediately following the transistor's announcement, research efforts at Bell Labs and in the industry as a whole focused on developing new types of transistors that would work better and more reliably than vacuum tube replacements. By the early 1950s, however, advances in semiconductor technology had pushed the transistor beyond its early aim to replace the vacuum tube. Transistors began to be designed that went well beyond the possibilities offered by tube technology, and the outlines of a distinct industry emerged for the first time.

2

From Tubes to Semiconductors

THE CONTEXT OF INNOVATION

World War II had fundamentally altered the field of electrical engineering, dramatically increasing its scope and changing its internal organization. Engineering colleges, particularly in the United States, had rapidly expanded during the war, and growth continued as nearly every institution launched a major effort to expand its research capabilities. In the field of electrical engineering, the increased importance of electronics in the military and civilian economies was reflected in the rapid growth of its professional associations, such as the Institute of Radio Engineers (IRE) in the United States and the Institute of Electrical Engineers in the UK. Further, as more jobs opened up in device research and manufacturing, special interest groups of these professional societies grew rapidly. The IRE's Professional Group on Vacuum Tubes, among others, sponsored a series of professional conferences at which nearly every milestone invention in the device field was first announced.

Electron devices after 1945 were the special beneficiary of many millions of the new multibillion-dollar U.S. research budget. The total annual expenditure by the federal government for research and development of military technology had already reached an unprecedented $2.6 billion by 1949. With the onset of the Korean War in 1950, President Harry Truman increased total military spending dramatically, from about $13 billion per year to about $50 billion. More than a little of this went directly to firms that developed or manufactured electronics. Funding for more basic research and development likewise increased, and within a decade it had risen to an astounding $12.4 billion, a figure that represented more than 2.5 percent of the gross national product. The space race, part science and part Cold War posturing, was one important source of research dollars and direct income. While the "rocket scientists" were the most visible heroes in this effort, hundreds of electrical engineers and physicists worked behind the scenes to develop the navigation and communication systems that made space travel and satellite communication possible. Vast new military communication, navigation, and defense networks were also being set up, such as the continent-spanning defense early warning (DEW) radar. The DEW line and other new military systems were often combinations of previously separate areas of technology, including radar, telecommunication, and computers. The performance and reliability of these systems were imperative under the threat of a nuclear war, a fact that had a substantial bearing on the way that new electronic components were being developed, which options were chosen, and which potential paths were not taken.

A substantial amount of this research effort was dedicated to developing smaller and smaller electronic components in order to make a broad range of military and scientific goals feasible. Space flight of the kind envisioned by U.S. and Soviet engineers demanded new control and communication techniques, many of which required highly miniaturized electronic devices to be practical. Miniature electronic components were also necessary for the development of advanced guided missiles and rockets, for communication satellites, and for developments in the field of aeronautics. The Minuteman missile developed in the late 1950s, for example, gave a significant boost to early semiconductor electronics, and it became one of the first early missile systems to use transistors. As a result, the Air Force provided funding for research on reliability improvement and miniaturization of transistors to the tune of at least $13 million, and about 800 transistorized Minutemen had been deployed by 1965.

Perhaps most importantly, the atmosphere of the Cold War bolstered the notion that research must remain a national priority. Projects that might not have been funded in the prewar period due to their lack of short-term commercial potential often became the recipient of substantial research grants during the 1950s, even ones with no immediate military applications. The result of this intense research activity was the appearance of a wide variety of new electronic devices. A few of these had immediate commercial appeal, but many survived only because of the hothouse environment of Cold War military spending.

While the cutting edge of electron device research had shifted to military and space applications, the civilian and consumer markets also drove considerable innovation. The biggest event was the reintroduction of television. First in the United States, then in the war-ravaged countries of Europe, black-and-white and later color television systems reappeared. In the United States, the major networks virtually abandoned their radio programming efforts, shifting their human talent to the new medium. Most other countries established government-sponsored networks to provide universal TV coverage. The result was a huge new market for television receivers, cameras, transmitters, and production equipment, all of which helped stimulate continued innovation in the device field. New consumer products such as tape recorders and transistor radios constituted smaller but still important markets.

Looming on the horizon also was the computer. Introduced during World War II, the computer rapidly moved from laboratory experiment to commercial enterprise. While at first reliant on established forms of electronic devices such as vacuum tubes, the computer over the course of the 1950s would begin to influence the direction of development in the field of transistors. Computers, then as now, came in many different forms, and the best remembered are the giant "mainframe" models made by the likes of International Business Machines (IBM). But smaller computers such as the ones used in missile navigational systems were sometimes the real drivers of technical change. This would become especially apparent in what was arguably the most exciting product of the decade, the integrated circuit.

VACUUM TUBES FOR COMMUNICATION

Vacuum tubes had been the fundamental building blocks of the electron device industry in the years before World War II. During the war, military applications drove the rapid development of vacuum tube technologies by creating a huge demand for advanced communications equipment and other tube applications. This resulted in innovations crucial to the war effort and led to fundamental advances in tube technology, including the development of the cavity Magnetron oscillator and new types of gas-filled tubes, both for use in radar technology. In the years immediately after World War II, vacuum tube technology continued to develop rapidly and remained a fundamental part of the rapidly expanding electron device industry. Tube technology was, in fact, commercially more important than semiconductor devices at least through the mid-1950s. Looking back at the history of the electron tube, one commentator in 1965 described its importance in the following way:

> It is a truism that almost every engineer's livelihood has been profoundly affected by the electron tube. Born with the invention of grid control by DeForest about 1906, it has been the mainstay of electronics for the past five decades. Radio, television, computers, sound motion pictures, and much of the field of energy conversion and automation, were either made possible or profoundly modified by outgrowths of the DeForest device. In addition, its influence has been felt in many other areas. The chemical industry, medicine and biology, transportation, education, finance—all have advanced and been altered by its use. ("The Future" 1965)

However, with the advent of the transistor and the rapid shift to solid-state semiconductor devices in certain applications, the character of the electron tube's influence began to change. Paralleling the growth of solid-state technology between 1950 and 1965 was a substantial change in the "product mix" of electronic tubes. Tube technology was displaced as the basic electronic building block of the electron device industry and instead shifted into what may be considered specialized niche applications that solid-state technology could not fill, such as high-power, high-frequency devices for use in communications equipment. As the electronics industry flourished in the years following World War II, these niche applications grew rapidly as well. Television picture tubes, microwave tubes, imaging devices, data-storage tubes, and other tubes that in some cases bore little resemblance to the original de Forest invention became increasingly important and maintained a relatively high dollar value into the mid-1960s (and in some cases until the end of the twentieth century). As a result, there was no appreciable decrease in the total dollar value of the tube manufacturing industry during this period, although the value of the tube industry relative to the electron device industry as a whole declined markedly. Thus, the character of the tube industry changed dramatically to reflect the impact of solid-state devices, while the size of the tube industry remained relatively stable within a larger, rapidly growing industry.

INTERNAL IMPROVEMENTS

Tube technology as a whole saw significant advances as tubes became smaller, more efficient, and able to handle greater "current densities" or power levels. Tubes were still unreliable due to the failure of elements inside the tube. In consumer electronics, tube failures were less costly to repair than in military or telecommunication equipment, so it was these latter applications that saw more intense efforts at improvement. One such effort, undertaken largely at Bell Labs, was in the area of cathode improvement. As early as 1930, the average life of cathodes in the tubes used in long-distance telephone "repeaters" (amplifiers) had been increased to 20,000 hours for low-current densities through methods such as improving manufacturing processes. Improvements in some tube designs dating to the 1930s allowed them to continue to be used successfully for many decades. For example, the Western Electric type 301A tube, used widely in repeaters, attained an expected working life of over fifty years. However, this required that the tubes were operated at currents and voltages far less than the maximum. In order to increase cathode life substantially, which was necessary for high-current-density applications to be economically feasible, researchers intensively investigated the characteristics of the metals, usually nickel or tungsten, used to make cathodes. Over the course of the 1950s, Bell Labs and others developed new ways of manufacturing cathode materials, and these allowed tube power levels to be increased.

NEW KLYSTRONS AND TRAVELING-WAVE TUBES

The traveling-wave tube was a new type of amplifier proposed in the 1930s and improved by Bell Laboratories researchers during World War II. After the war's end, researchers identified the wide-band, high-gain characteristics of the tube as potentially valuable in the creation of new communications systems, including satellite relays and ground-based microwave radio communications. AT&T was at that time preparing to employ microwave relays as the basis for a new system of long-distance telephony. The system would essentially consist of numerous amplifier/relay stations and their associated antennas and towers, strung across the countryside along high-traffic telephone routes. These towers are commonly seen today, often along highways, and are identifiable by their peculiar, horn-shaped microwave antennas. The Western Electric 444A tube, used in the system's main amplifier, became the first traveling-wave tube to be produced in significant quantities (claimed to be about 20,000 by the early 1970s).

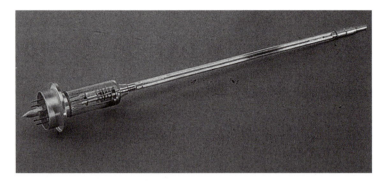

Bell Telephone Laboratories traveling-wave tube, 1967. Courtesy Lucent Technologies Inc.

THE FIRST TRANSATLANTIC TELEPHONE CABLE

It may come as a surprise that telephone communication across the Atlantic Ocean was only possible by radio before World War II, as compared to transatlantic telegraphy, which had been in use since the nineteenth century. The main problem was amplification. Because telephone signals required amplification at regular intervals along the cable route, it was necessary to build "relays" that were waterproof and highly reliable. Work on developing these types of tubes began in the 1930s, and a short submarine cable was laid for the first time in 1950, between Key West, Florida, and Havana, Cuba. The transatlantic cable project, however, was likened by Bell Labs president William O. Baker to being as difficult as the feat some years later of putting the first satellite into orbit. Longevity testing of the critical cathode structures had begun in World War II, and other careful performance tests were conducted on filaments and grids. The tubes themselves were assembled at Bell Laboratories, rather than AT&T's regular manufacturing arm, Western Electric. After being constructed in "clean-room" conditions by workers in smocks and gloves, the tubes were tested in-circuit for 5,000 hours before being selected for use. The attention given to the design and manufacture of these tubes was proven when the cable was installed in 1956. When it was finally made obsolete by satellite technology in 1978 and taken out of service, the more than 300 tubes had performed continuously for twenty-two years without a failure. Similar performance was obtained from the type 455A-F tube developed in 1955 at Bell Labs for a new landline cable communication system. Coaxial cable, similar to that used in homes to deliver television programming, was initially developed for long-distance telephone communication. Bell Labs applied the same stringent reliability requirements on this tube, and as a result only two tubes out of over 5,800 in use had failed after a total of over 700 million cumulative hours of use.

IMAGING AND DISPLAY TECHNOLOGIES

In many ways the field of imaging and display devices after World War II is a microcosm of broader historical trends in the electronics field: the 1950s saw the maturation of electron tubes, followed by the emergence of semiconductor replacements, followed by the incorporation of the microprocessor. As in the field of communication, tubes remained in use in certain applications, some of which remain very important to this day. A number of important display technologies, such as those involving liquid crystals, have also been developed that are only tangentially related to the larger story of tubes and semiconductors, and as a result the field of electron imaging and display devices as a whole is quite complex and difficult to summarize.

To begin, it should be pointed out that a wide variety of simple display devices were developed in the years after World War II that made use of tube technology. Many of these were in the field of instrumentation and computers. Engineers desired numerical, alphabetic, or other displays for a large variety of test equipment and other electronic applications. These demands soon outstripped the speed and flexibility possible with older forms of electromechanical devices such as the totalisator or "tote board" widely used to display racetrack results, election information, or train arrivals. One of the most popular of the postwar electronic display devices, for example, was the Nixie tube, also known as the "numicator." The Nixie tube, invented in 1954, consisted of an outer mesh anode and ten wire cathodes shaped to form numbers. Electronic switching circuits connected to the Nixie tube caused any one of the ten cathodes to light, so that by installing a row of Nixies, one could display a multidigit numerical output. If necessary, the numbers could be switched rapidly to allow the monitoring of a fluctuating or varying event, such as time. It was a relatively simple device, and was used for many years in various applications including calculators and test instruments.

CATHODE RAY TUBE IMPROVEMENTS

In the period between the eclipse of the "A-scope" and the early 1950s, radar CRTs were little more than standard television picture tubes with P7 phosphor screens. Beginning in the early 1950s, however, radar tubes began to take on more of their own identity as specialized tubes and were designed to overcome the problem of image retention. In 1953, for example, RCA developed early versions of the direct storage CRT. These tubes used the phenomenon of secondary electron emission to retain images longer than possible with simple phosphor emission. Secondary electron emission is a result of the fact that many secondary electrons are emitted from the phosphor when electrons strike it at high levels of acceleration. At a certain point, however, these electrons begin to be trapped in the phosphor due to a phenomenon known as "sticking potential," and the proportion of emitted electrons goes down as the acceleration goes up. By employing both an electron gun to write the image desired and a "flood gun" to flood the screen with electrons, the sticking potential can be used to create images that are retained for long periods of time. These and other types of direct storage tubes were produced commercially for radar and oscilloscopes beginning in the mid–1950s by companies including RCA, Hughes Aircraft, and IT&T. Other important advances during the 1950s in radar and oscilloscope CRT technology include the development of the monoaccelerator CRT in 1954 and the 1953 invention of the spiral accelerator, in which the acceleration of the electron beam could be smoothly adjusted through the application of a gradually increasing accelerating voltage. This helped eliminate the beam distortions that plagued other CRTs and became one of the key advancements in improving oscilloscope design.

To the general public, television was one of the most visible "benefits" of military research during World War II, and RCA and others quickly made the reconversion to TV production as soon as the war ended. Wartime techniques developed at RCA for the manufacture of radar CRTs, and the use of automatic or semiautomatic assembly lines, were applied to civilian television manufacturer after the war and greatly reduced the cost of TV receivers. The number of television stations on the air expanded rapidly once the war was over, and in 1951 television was transmitted coast-to-coast in the United States for the first time. Yet all this had taken much longer than engineers originally hoped, and proved to be considerably more expensive than originally envisioned. David Sarnoff, president of RCA, later remembered that Zworykin had expected television to be much easier. "I asked him how much it would cost to develop TV," Sarnoff recalled. "He told me $100,000, but we spent $50 million before we got a penny back from it." Despite this, commercial television developed rapidly throughout the 1950s and quickly became an important part of American culture.

Part of the commercial appeal of television in the 1950s came through developments in television CRT technology. One important advance was the ability to make CRTs larger, more rectangular (in part to correspond to the rectangular shape of motion picture film, already a major source of TV content), brighter, and flatter so that television would be more pleasant to watch. Early television CRTs were all round and masked the edges of the tube so that the viewer saw only the central rectangle. Only in 1949 did Hytron of New York and others begin to blow tubes in rectangular molds to match the displayed images. General Electric soon followed with its own rectangle-faced tube, and by the early 1950s the rectangular tube had fully replaced the round tube. Tube size increased rapidly as well: as late as 1947 most televisions had screens that were 7 inches diagonal or smaller, but by the early 1950s screen sizes had tripled, and 20- and 21-inch rectangular television CRTs rapidly became standard. Other innovations included the introduction of gray shades to enhance screen contrast and a shift from 70-degree deflection angles to 90-degree deflection angles, an important change as screen size increased weight and size became more of a factor; both were reduced by this measure. Westinghouse Corporation was the first to shift to a 90-degree deflection angle with its 21AMP4 line, and many other companies soon followed suit with various screen sizes. Price competition drove CRT prices down during the 1950s, and this in turn compelled manufacturers to introduce cost-cutting design changes including the introduction of lower current filament circuits (which allowed a smaller, cheaper power supply circuit) or even the elimination of the power transformer altogether through the use of "series-filament tubes." This measure, already in use since the 1930s for consumer radio receivers, demanded that the entire complement of tubes for a television be designed together. Their filaments were then wired together, and the entire line voltage of 110 volts (220 in Europe) could be applied directly to the tubes without the need for a voltage-lowering transformer. While this occasionally resulted in a spectacular failure (or fire), and increased the risk of accidental electrocution, it dramatically reduced the cost and often the weight of the TV receiver.

A major improvement in CRT technology was the development of the commercial shadow-mask tube around 1950. By this time, researchers were convinced that for sharp, moving images, the CRT phosphors needed to be applied as a matrix of "pixels" corresponding exactly to the same number of pixels in the original camera tube, and they had spend considerable time settling on how many pixels the screen should hold. But focusing the beam so that electrons struck only the appropriate pixel was a problem. Harold Law and others at RCA placed a perforated metal mask near the screen, between it and the electron gun. The effect was to cover the area around each pixel and block electrons that had strayed off their path. The shadow-mask idea, greatly elaborated, remained in use at the end of the twentieth century.

Considerable effort was also spent during the 1950s to perfect a color CRT and to develop standards for all-electronic color television. Color television had been demonstrated as early as 1929 at Bell Labs, but this experimental proof of concept transmitted only 50 lines of information as compared to the 525 lines adopted for postwar TV. RCA demonstrated the shadow-mask "triniscope" color tube to the FCC in 1950 as part of its campaign to gain acceptance for its color television technology. The color shadow-mask tube used three electron guns aimed at clusters of tiny red, green, and blue phosphor dots applied to the inside surface of the CRT. As in the monochrome shadow-mask tube, the mesh mask was aligned near the screen so that a small hole was immediately behind each RGB dot cluster. The system as demonstrated performed poorly and the receivers were very large, which may have been a factor in the FCC's rejection of RCA's "all-electronic" color television in favor of a competing system from CBS. In the CBS system, color was achieved through the use of a black-and-white CRT with a large, rotating color wheel with red, green, and blue translucent panels. The eye's persistence of vision made the rapid sequence of red, green, and blue images combine into one full-color image.

RCA as well as Philco, the Hazeltine Corporation, General Electric, and many others worked feverishly in 1950 to develop a system of color television that would be compatible with existing black-and-white receivers, believing (correctly) that the FCC would insist on this. By 1951 representatives from these firms pooled their resources and worked together to establish standards for this compatible color system. The FCC adopted the resulting standards on December 17, 1953, opening the way for commercial color television. Surprisingly, however, color television sales did not increase as rapidly as expected, and color television did not rival black-and-white television in significance until as late as 1968. In the interim, Japanese firms entered the monochrome television market and gradually took market share away from U.S. and European firms. Engineer Masaru Ibuka, working with the Sony Corporation, developed an improved color tube called the Trinitron, which Sony demonstrated in 1968. Two years earlier, Sony had been offered a license to a General Electric invention, a color tube with three electron guns placed in-line rather than in a triangle configuration. Sony rejected the offer, but developed the Trinitron along these same lines. In addition to the improved electron gun, the Trinitron offered a sharper and brighter picture through improved focusing of the beam through a single, large-aperture lens, and the use of an "aperture grill." This grill, consisting of a metal sheet with long, thin slots rather than the round holes of the shadow-mask tube, more accurately focused electrons onto phosphors that were arranged as side-by-side vertical stripes rather than clusters of three dots each. The Trinitron was probably the last major improvement to the color CRT for television, although numerous incremental advances came before CRT production began to slip relative to the production of other types of displays in the early twenty-first century.

IMAGING

Television's emergence in the late 1940s and early 1950s also drove the development of devices to convert visual images to electrical signals. The first television transmissions were nearly all "live," using images supplied by a variety of devices. Vladimir Zworykin's original Iconoscope camera tube was largely superceded in 1943 when Albert Rose, Paul Weimer, and Harold Law developed the Image Orthicon, which produced relatively high-resolution pictures and produced a better quality of images. The next major advance was the Vidicon tube, introduced by RCA's Paul Weimer, Stanley Forgue, and Robert Goodrich in 1950. The Vidicon was the first camera tube to employ the principle of photoconduction rather than photoemission to generate the video signal. It used a light-sensitive film placed over a signal electrode to form a photosensitive target. An electron beam then scanned the film and provided an output signal directly to the output. As a result, it was much smaller and more practical than either the Iconoscope or the Image Orthicon, making applications such as portable cameras more practical. A wide variety of photoconductive tubes was soon developed using the basic principle of the original Vidicon. These have been developed and identified under different trade names, but Vidicon has become the generic name for all such devices.

Other technologies developed during the late 1940s and 1950s that would eventually have important applications were phototransistors and photodiodes. In 1949, not long after the first transistor was developed, John N. Shive of Bell Labs demonstrated the first working phototransistor, using a germanium point-contact transistor. The device was in essence a point-contact transistor without an emitter lead. Light striking the device performed the emitter function, resulting in a small current that appeared at the collector. The device was used experimentally in a Bell System device called "card translator," used as part of the telephone system. In it, information was stored on punched paper cards, and light shining through the cards was detected by the phototransistor. Although hardly a device for collecting complex images, the phototransistor concept would eventually emerge as a television and video-imaging technology in later years. Further, although this early application of the phototransistor remains an obscure event in the history of technology, it was apparently the first commercial application of transistors in the Bell system, and probably the first civilian application of any kind for the transistor.

OTHER DISPLAY TECHNOLOGIES

Although the CRT was the most commercially important display device in the 1950s, other types of technologies were developed or investigated during this time, some of which would prove extremely important in later years. One such development was the light-emitting diode (LED). Light-emitting diodes had their roots in the experiments of an engineer named H. J. Round in 1907, who electrically stimulated silicon carbide into electroluminescence. The principle of stimulated emission was formulated by Albert Einstein in 1917, and inspired considerable experimentation around the world, much of it aimed at microwave devices. In fact, this work was linked to the invention of the maser and later the laser. By the mid-1950s, there was considerable interest among electrical engineers in p-n junction semiconductor devices that could operate at high frequencies. In 1955 R. Braunstein noticed that infrared radiation was produced by carrier injection in gallium arsenide and indium phosphide semiconductors. He proposed that this was the result of the recombination of pairs of electron holes, and researchers soon recognized that a variety of semiconductor compounds made from elements in columns III and V of the periodic table are able to generate light. Gallium phosphide (GaP) P-N junction semiconductors were the most promising since they could be used to generate light in both the red and green parts of the visible spectrum. In 1957 British physicists J. W. Allen and P. E. Gibbons built the first point-contact gallium phosphide light-emitting diodes. Substantial research on similar devices was subsequently undertaken during the late 1950s and early 1960s by a variety of researchers, laying the groundwork for important later developments in LED technology.

TRANSISTOR IMPROVEMENTS

Following the announcement of the transistor in 1948, Bell Labs moved quickly to commercialize it, but it took time to learn how to manufacture reliable transistors in quantity. In the meantime, in 1950 Shockley published his important book detailing what he and his colleagues had learned from more than a decade of research. Entitled *Electrons and Holes in Semiconductors*, the book became an instant standard.

Work on the junction transistor, which had been languishing during most of 1950, was revived later in the year after William Shockley became convinced that there was an important military application for it. So-called proximity fuses, miniaturized radar sets placed inside missiles or bombs to detonate in the proximity of a target rather than on impact, had been successfully demonstrated in World War II using vacuum tube electronics. Some of these fuses were placed in artillery shells, where they were subject to extreme stresses when the shell was fired. Shockley believed that the rugged junction transistor would be a superior replacement for the vacuum tube. Morgan Sparks of Bell Labs succeeded in pulling "double-doped" crystals with very thin junction layers in early 1951; the thinness of the layers (among other factors) determined the high-frequency characteristics of the device. The tiny devices also consumed even less power than their point-contact cousins, and introduced less "noise" to a circuit, making them better

In a controversial but often-published press release photo, William Shockley examines the first transistor, while colleagues Walter Brattain and John Bardeen look on. Shockley was later criticized for attempting to "steal" credit for the device, which was later shown to be largely the work of the other two men. Courtesy Lucent Technologies Inc.

for use as highly sensitive, low-power amplifiers. The importance of the junction transistor cannot be overstated in the story of electron devices: although the production of the original type of point-contact transistors would briefly sustain the developing semiconductor industry, the future of the transistor lay with such crystals rather than devices assembled from discrete parts. "The point contact transistor," historians Braun and MacDonald have written, "with its delicately positioned electrodes, harked back to the triode valve and to the cat's whisker rectifier; the junction transistor, in which action took place within the body of the semiconductor, pointed the way to modern solid-state electronics" (Braun and Macdonald 1978, 43).

Tied to the success of the junction transistor was a new way to process germanium crystals. Between 1950 and 1951, Bell Labs researcher William Pfann discovered an important new way to improve the quality of the germanium crystals being made in the lab. He called the process "zone refining," and it was essentially a way to temporarily heat a germanium rod by moving a special heating ring along its length. As the rod was passed through the ring, a part of the crystal briefly melted and then recrystalized. This tended to drive the impurities to one end of the rod, which could then be discarded. Multiple sweeps through the ring resulted in extremely pure germanium. Kept secret for two years, zone refining was finally announced in 1952.

THE TRANSISTOR SYMPOSIA

Bell Laboratories decided (or perhaps was compelled) to follow an active policy of public disclosure regarding the research and development of transistors. The Department of Defense wanted AT&T to disclose details of the device to its contractors, who were clamoring for more information. Further, AT&T's management believed that disseminating the transistor would lead to its greater acceptance as a replacement for tubes in consumer, industrial, and military applications, generating patent licensing revenue along the way.

Beginning in September 1951, Bell Labs held symposia at its Murray Hill, New Jersey, facility, where engineers explained how to make point-contact transistors and revealed some of Bell's progress in the area of junction transistors. The first symposium did not cover the new junction transistors very thoroughly, and was notably silent about the manufacturing processes involved and their application to military systems. More than 300 people (mostly military personnel) attended the first symposium, each paying a $25,000 entrance fee. Jack Morton, then a senior manager of the semiconductor area at Bell Labs, gave an excited talk about the potential of transistors and made bold claims about Western Electric's prowess in their manufacture and Bell Labs' mastery of their design. This helped disseminate the results of transistor research to the electrical engineering community as a whole, although some attendees objected to being denied information about manufacturing. Some companies intended to enter transistor manufacturing themselves rather than purchase Western Electric transistors. For the time being, these firms had to reverse engineer the manufacturing process. Many were successful in doing so. The staff of the Philips company of

Jack Morton. Courtesy Lucent Technologies Inc.

Eindhoven, the Netherlands, in fact, was able to build their own transistor without even attending the seminar, building the device by relying solely on descriptions taken from U.S. newspapers. While AT&T did not necessarily encourage others to make transistors, it did not actively attempt to prevent it. Then, after September 1951 the company began granting nonexclusive patent licenses to manufacture transistors at a relatively low royalty rate. Among the first companies to take these $25,000 licenses are a few that are still in business today, including Texas Instruments, International Business Machines (IBM), Hewlett-Packard, and Motorola. Licensees were invited to a second symposium in April 1952, where the secrets of making transistors were finally revealed. These licensees had considerable success manufacturing and selling transistors in the early 1950s, and with several other firms formed the core of a rapidly growing new industry.

Jack Morton

Jack Morton joined Bell Labs in 1936 as a vacuum tube engineer, and by the 1940s was moving into management. Following the invention of the transistor, Bell Labs split the transistor research team into two parts, one focused on semiconductor physics and the other on the production of transistors. Morton was put in head of the production group. Under his direction, Bell Laboratories set up a small manufacturing facility to turn out the first commercial transistors (called simply "Type A"). However, Morton pushed hard to abandon the early type of point-contact transistors and adopt junction devices. By the early 1950s, he was also an advocate of the idea of disseminating transistor knowledge, and helped lead the famous Bell Labs Symposia on the subject. In the mid-1950s, he also became the outspoken champion of silicon as a device material, rather than the more established germanium. Although other firms such as Texas Instruments are often credited with pioneering silicon technology, their interest in it was partly due to Morton's persistence.

Interviewed by *Business Week* in 1961, Morton stated that solid-state electronics would become the largest industry in America, overshadowing even the steel and automobile industries. Morton in these years became something of a national figure, owing to his success in semiconductors and to his prominent position at Bell Labs, then the world's leading industrial research and development laboratory. Along the way, he had made a number of enemies. His compulsive personality and persistent nature rubbed many the wrong way, and although AT&T management appreciated his ability to get things done, many of his colleagues and employees privately disliked him. Suddenly, though, in late 1971 he was gone. He was found dead in his car by the side of a road near Neshanic Station, New Jersey, his body on fire. He had been murdered, apparently after an argument at a bar, by a man who was later convicted of the crime.

NEW TYPES OF TRANSISTORS AND NEW WAYS OF MAKING THEM

Western Electric meanwhile began manufacturing the first point-contact transistors for sale in 1951. By April 1952, Western Electric was manufacturing about 8,400 of these transistors per month, while smaller numbers were being produced by companies such as Raytheon, RCA, and General Electric. Western Electric was also beginning to produce junction transistors for commercial sale, although at the slow rate of less than 100 per month. But Western Electric transistors were barely into production when engineers began to announce new types of devices, the numbers of which grew during the next few years. In 1952, for example, General Electric announced that John Saby had developed a method of alloying indium dots to opposite sides of a thin germanium sheet to produce a new "alloy junction" transistor. The alloy junction transistor was able to operate at higher frequencies and currents than earlier junction transistors, but required an extremely thin layer of germanium that was difficult to manufacture accurately. Nonetheless, RCA was able to take the alloy junction transistor and put it into production rather quickly, introducing it as a competitor to Western Electric's junction transistors.

Charles W. Mueller: On the Difficulties of Staying Clean

Charles Mueller was an engineer at RCA who contributed to the development of the alloy junction transistor.

[We] went up to the tube plant in Harrison [New Jersey] to set up a group to make transistors. . . . We finally got a corner in the laboratory there where we set up our line, which consisted of ten girls doing the job. We wanted things very clean, so they said, Ok, they would clean the place for us. They got in the janitors and they went through the ducts which were very big, about four or five feet in diameter, and they scraped the dust off the walls. But then, of course, they never really got it out of the pipes for two months. It just kept draining out of the pipes. Some of it was so big that it was hard to tell the difference between it and the indium dots. Ethel Monan would look at the shape. If it was round, it was indium and if it was a particular shape, it was dust. Finally we got that dust out of there. The building's supervisor was located in a building three blocks away and to him we were just another room to clean. We really wanted to get better cleaning. After some negotiations, he said he would give us another janitor. Fortunately, I came in one day when this janitor was at work. We were in one corner of the laboratory and he used the push-broom to sweep the dust from the whole laboratory down into this corner. Every time he tapped the broom, huge volumes of dust went into the air. Then he would take the dust and dump into a can in our corner of the room. Nobody told him in which direction to sweep the dust or anything like that. We wanted him to use the vacuum cleaner, but the people that did the cleaning said that this did not fit into their budget, that they couldn't do anything like that. We finally bought a vacuum cleaner on the engineering budget.

Source: Charles W. Mueller, electrical engineer, an oral history conducted in 1975 by Mark Heyer and Al Pinsky, IEEE History Center, Rutgers University, New Brunswick, New Jersey.

The next year, the Philco Corporation entered the transistor market with its own design, called the surface-barrier type, a modified form of the alloy junction transistor. This type of transistor was created by using a technique called jet etching to spray an electrolyte on both sides of a germanium wafer in order to make it extremely thin. An alloy of indium sulfate was then applied to the wafer to form the junctions, and then metal contacts were put in place in the appropriate spots. These contacts were then heated until they melted and formed junctions. Unfortunately, the surface-barrier transistor was quite fragile due to the thinness of the germanium wafer and as a result was of limited usefulness.

Herbert Kroemer, a German physicist working in the United States, in the 1950s began working on a class of semiconductor devices incorporating what are today called heterostructures. A heterojunction, for example, is defined as a junction between two semiconductors with widely differing band gaps. Heterojunction research, stimulated by Kroemer's publications on the

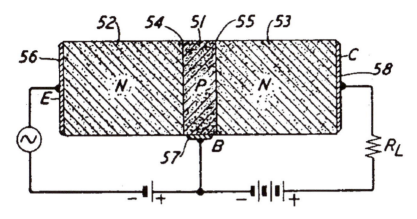

William Shockley's junction transistor of 1948 became the basis of nearly all subsequent efforts in the semiconductor field. The transistor was a sandwich of n-type and p-type germanium layers. A small signal (shown symbolically below E) applied between the middle "base" layer (B) and the n-type emitter layer (E) allowed a larger current to flow in a circuit consisting of battery (drawn symbolically below and to the right of B), output (RL; here a resistor is substituted), the emitter layer, and the collector layer (C). The second, smaller battery below and left of B provides the "bias" current, which in effect prepares the base for action. U.S. Patent 2569347.

subject beginning in 1954, would lead in all sorts of different directions. Kroemer himself continued his research, publishing important theoretical papers while working at RCA Laboratories in Princeton, New Jersey. Later, in 1963 while working for Varian Associates, he published the results of his research on the groundbreaking double heterostructure semiconductor laser. However, few of these ideas could be commercialized at the time. Heterojunction lasers, for example, only came into wide use after the early 1980s.

Rapid innovation in design continued and came to be one of the primary characteristics of the semiconductor industry as a whole. By 1953 there were sixty distinct types of transistors being manufactured. By 1957 this number had jumped to at least 600. While most of these were minor variations on the basic theme, it is indicative of the widespread interest in transistors, their adaptability to various uses, and the proliferation of transistor manufacturers.

THE EMERGING TRANSISTOR INDUSTRY

By the mid-1950s the production of germanium transistors had nearly replaced point-contact production and formed the basis of an expanding industry in the United States. In 1951 there had been just four domestic companies making transistors commercially. By 1952 there were eight, by 1953 fifteen, and by 1956 there were at least twenty-six companies producing over $14 million worth of germanium transistors per year. This rapid growth marked more than simply the emergence of a new industry. It also marked the beginning of a new era in electronics. As the number and types of transistors available increased, and as advances in semiconductor technology pushed the transistor past mere equivalency with vacuum tubes, semiconductors began to be much more than just an efficient replacement for tubes. Engineers began to design transistorized devices that did things previously thought impractical or impossible. Some of these applications were designed for the individual consumer: Texas Instruments, for example, manufactured the transistors for the first consumer transistorized pocket radios in 1954.

However, despite the fanfare generated by the transistor radio, commercial success was slow in coming in the 1950s. Instead, the U.S. military was by far the single most important consumer of transistors. The continued development of radar and aircraft electronics, and the advent of guided missiles, spurred the military to seek out reliable, low-power, miniaturized electronic components, and as early as 1952 non–Bell Labs suppliers of semiconductor devices had signed over $5 million in military contracts. One of the first military applications for Western Electric's point-contact transistor was the so-called AN/TSQ series of specialized digital communication devices. These "black boxes" took in analog data from radar sets, converted it to digital form, and transmitted it over telephone lines to another black box, where it was decoded and sent to a radar display screen. These systems became an integral part of the Nike missile systems that were installed in numerous places around the United States. By 1954, Bell Labs would build the first transistor-based computer, the transistorized digital computer (TRADIC) for a military sponsor, the U.S. Air Force. Bell Labs itself had a close relationship with the military, having signed a joint services contract as early as 1949 that allowed it to undertake specific research related to transistors. Nearly half of the incoming transistor research money at Bell Labs in the early 1950s came from military sources, and the Army Signal Corps underwrote the construction or improvement of transistor production lines for Western Electric, RCA, Raytheon, and Sylvania. The close relationship between the military and the semiconductor industry would continue in the years to come, and would prove crucial to the development of the industry as a whole. As one observer of the industry later recalled, "I can hardly think of a single company [in the United States] in the fifties that did not enjoy significant government support for their semiconductor operation. . . . I can absolutely assure you that we would never have enjoyed the success that we enjoyed in that period had we not had the government money" (Braun and McDonald 1978, 72).

The value of military sponsorship grew explosively in the 1950s, but the future significance of the military market for semiconductor development may not have been obvious to people at the time. From the perspective of the early 1950s, the transistor was noisy when compared to tube technology. It could handle less power than a tube, it was more likely to suffer damage from sudden power-supply fluctuations, its characteristics changed dramatically with changes in temperature, and it had a very narrow bandwidth. How would the high-power, ultrahigh-frequency requirements of military radar and microwave communication systems be satisfied by this anemic little device? Furthermore, due to the laborious manufacturing process needed to make usable transistors, it was difficult to produce two devices with exactly the same characteristics, and early transistors tended to be quite unreliable. They were also very expensive compared to tubes—$20 for the earliest models, and still about $8 in late 1953—whereas a comparable tube was as little as a dollar. Transistors, in short, were in many ways an unattractive product in the early 1950s. Nonetheless, the use of the transistor began to expand, especially for applications where its price was less important than its technical advantages. The first nonmilitary use of the transistor outside the Bell system was in hearing aids. AT&T gave hearing aid manufacturers royalty-free licenses in honor of Alexander Graham Bell's work with the deaf. Sonotone Corporation, a well-known hearing aid maker, was the first to offer a transistor hearing aid in 1952. The tiny device's fame was spreading by then, as *Fortune* magazine called 1953 "The Year of the Transistor."

GERMANIUM TO SILICON

Even though other types of semiconductors were known, germanium was the basis of all production transistors from the late 1940s to the early 1950s. The industry as a whole was quite surprised when Gordon Teal at the virtually unknown Texas Instruments Corporation announced that he had succeeded in making a silicon transistor in 1954. Teal had in 1951 worked at Bell Labs with Ernie Buehler and had found a way, in the laboratory at least, to create silicon crystals and form p-n junctions in them. He left Bell Telephone Laboratories in 1952 to return to his home state of Texas and head up Texas Instruments' growing transistor development department. There, he continued his work in silicon, because he recognized that there was a critical disadvantage to the germanium transistor; as it

Gordon Teal, whose work at Bell Labs on purifying germanium contributed to the improvement of transistors, is shown in this 1951 photo. Courtesy Lucent Technologies Inc.

heated up, it gradually stopped working until it reached 167 degrees Fahrenheit, where it quit completely. When temperature could be controlled, germanium was fine, but it could not be relied upon in the battlefield or, indeed, in many types of electronic equipment.

Gordon Teal on Crystal Pulling

Gordon Teal was a researcher at Bell Telephone Laboratories in the 1940s, where he helped develop methods for "growing" semiconductor crystals.

Late one afternoon around quitting time, I encountered John Little, and we got to talking about our work. He started by telling me how he needed a germanium rod small enough in diameter to be cut by a very small wheel in order to minimize waste. I could see that here was an opportunity to make a rod for someone who had a real job to do.

As we were getting on the bus for Summit, New Jersey, I said, "Sure, I can make you a rod by pulling one out of a germanium melt. And, incidentally, it will be a single crystal, too." As soon as we got on the bus, we started sketching. All we needed was something that would pull the rod out smoothly and would withstand the heat. . . . By the end of the three-mile ride into Summit, we had sketched the equipment, and two days later, on October 1, 1948, we completed our crude machine in John's New York City lab. There we pulled our first single crystals of germanium. We did this without getting anyone's permission or approval and acted only on our own personal ideas.

Source: Gordon K. Teal, an oral history conducted on December 17–20, 1991, by Andrew Goldstein, IEEE History Center, Rutgers University, New Brunswick, New Jersey.

Teal and his team, including physicist Willis Adcock, accomplished his feat by rejecting the newer methods of making transistors, such as GE's alloying technique, and turning back to the methods of pulling crystals out of molten vats of material, a technique that he had used at Bell Labs. Silicon, unfortunately, is a more difficult material to work with than germanium, having a melting point so high that it tended to absorb all sorts of impurities from the atmosphere and the furnace. They finally gave up making their own silicon and purchased a purified version of it from the DuPont Corporation. With this starting material and two years of effort, Teal and his team finally succeeded in creating what he called "grown junctions" (to distinguish them from the junctions now being made by alloying) in a crystal in early 1954. Their problems did not end there. Silicon's physical properties also meant that the devices would have to be very small, so that when the metal electrodes were bonded to the semiconductor material, they would have to be very closely spaced.

Building a transistor with this level of precision created new manufacturing problems. However, the first silicon transistors that Texas Instruments offered could operate at higher frequencies and higher temperatures than the germanium transistors being produced at the time. Teal rushed to produce a few prototypes and to find an academic conference where he could announce the breakthrough. At the Airborne Electronics conference sponsored by the Institute of Radio Engineers, Teal staged a dramatic demonstration of the new silicon transistor. Instantly, although for only a short time, Texas Instruments became the sole manufacturer of the world's high-performance semiconductor devices. It took other manufacturers some time to make the switch to silicon, but by 1956, silicon alloy diodes (made by virtually the same processes as transistors) were also being manufactured by fifteen companies in the United States and a half dozen or so in other countries.

VAPOR DIFFUSION

Despite its promise, substantial problems kept others in the semiconductor industry from immediately embracing the use of silicon. Contaminants crept in from all sources, and even the technique of zone refining was not as effective on silicon as it was on germanium. As early as 1947, however, Russell Ohl, Jack H. Scaff, and Henry C. Theuerer at Bell Labs had known of a technique known as vapor diffusion that would become useful for addressing this problem. In vapor diffusion, the melted semiconductor material is exposed to the desired impurity in the form of a vapor. The impurity tends to diffuse into the surface of the semiconductor, resulting in a junction similar to that obtainable by other methods of manufacture. This process avoids the need to manually add doping pellets to the crucible containing the liquid silicon or to alloy them onto the surface. By regulating the length of time of this diffusion and the temperature at which it takes place, the depth and density of penetration of the impurity into the semiconductor material can be controlled with great accuracy.

It was not until 1954 that Bell Labs announced its first diffused-base product, and ironically it was not a transistor or an ordinary diode, but a solar cell. Daryl Chapin, Calvin Fuller, and Gerald Pearson invented the silicon solar cell over the course of searching for a replacement for batteries used to power telephone equipment in tropical locations. Selenium solar cells on the market converted perhaps 0.5 percent of the sun's energy to electricity, but Pearson discovered that a boron-doped silicon rectifier converted up to ten times that much energy. This diode consisted of a large, thin slice of silicon into which a junction had been formed through diffusion. Soon, AT&T was employing small panels made from these cells as "solar batteries" to power pole-mounted telephone equipment in rural areas around the United States and elsewhere. The first was installed in the small town of Americus, Georgia, in 1955.

By late 1954, diffused-base transistors were also being built. Bell Labs researcher Charles A. Lee diffused a layer of arsenic into a p-type germanium semiconductor to form a base layer. He then alloyed an emitter layer from evaporated aluminum onto the base region of the germanium to form the transistor. In the laboratory, the new germanium device was capable of working at 500 MHz. Early the next year, Morris Tanenbaum at Bell Labs produced the first diffused silicon transistor. Tanenbaum diffused antinomy and aluminum simultaneously into a silicon wafer, resulting in an n-p-n structure that could operate at a frequency as high as 120 MHz—the highest of any type of transistor.

Engineers at Bell considered vapor diffusion so important that they held a second series of symposia in 1956 to disseminate the process to the rest of the industry. By now several new companies had entered the transistor and diode business, such as Fairchild Semiconductor and Shockley Semiconductor, a new firm created by William Shockley. However, the potential advantages of vapor diffusion initially provoked little enthusiasm, since the rest of the industry was at that point based on the manufacture of germanium alloy semiconductors. In retrospect, however, the reasons for Bell's excitement are easy to see. Transistors produced through vapor diffusion achieved better frequency performance and were more reliable than transistors produced through other methods available at the time, making them desirable for military applications such as data processing that required high levels of reliability. More importantly, vapor diffusion allowed batch processing for part of the manufacturing process, pointing to the possibility of higher quantity production. As a result of these advantages, and with substantial support from the military, Bell Labs decided to focus its research efforts on vapor-diffused germanium and silicon transistors rather than on grown-junction and point-contact transistors. By the end of the decade, Western Electric had begun to market highly reliable diffused-germanium transistors on a commercial basis and had made substantial advances toward producing commercially viable diffused-silicon transistors as well.

LEO ESAKI AND THE TUNNEL DIODE

The tunnel diode was first discovered in 1958 by Leo Esaki, a brilliant Japanese Ph.D. student working for the Sony Corporation. Esaki was investigating the properties of heavily doped germanium p-n junctions for use with high-speed bipolar transistors. He discovered a device that conforms to one of the predictions of quantum mechanics, that electrons will "tunnel" through the energy barriers posed by p-n junctions under certain circumstances. The diode then became, in effect, an amplifier, although physicists preferred to call the phenomenon "negative resistance." The tunnel diode effect was subsequently demonstrated in a number of other materials, including gallium arsenide. Scientists outside of Japan quickly recognized the significance of the device, and in June 1958 Esaki was invited to deliver an address on tunnel diodes at the International Conference on Solid State Physics in Brussels. Esaki later recalled that he expected only a brief introduction from the session's moderator, William Shockley, but was surprised to hear Shockley heaping praise on his work on this promising new high-frequency device. This attention propelled the surprised graduate student and his tunnel diode to fame, and in 1973 Esaki was awarded the Nobel Prize in physics. At this time, Sony was still a fairly small company, and it was a breakthrough for them to produce a Nobel laureate. In a bemused look back at the event, Esaki wrote that "a company with 500 employees cannot afford to pay out that much in R&D expenses, so I think I earned the cheapest Nobel Prize in history" (Esaki 2000).

The tunnel diode's unusual characteristics made it commercially useful as a microwave oscillator, since it operated at high frequencies and was smaller than comparable transistors. By 1959 RCA was producing commercially available 1 GHz tunnel diodes, and in the early 1960s tunnel diodes began to be used in a variety of applications where transistors did not operate as well. Esaki diodes, for example, were widely used in UHF and microwave applications throughout the decade.

The high switching speeds possible with tunnel diodes also led many researchers to believe that the devices would be valuable in computer circuits. However, that promise has not been fulfilled. While capable of higher switching speeds than transistors, engineers found it uneconomical to manufacture the devices and to design switching circuits for diodes rather than transistors. Charles Mueller, one of the leading researchers at RCA who worked on Esaki diodes, recalled later, "You can make many circuits that work, in fact people made complete television sets using only the tunnel diodes. At first it was thought that these were going to be very cheap to make, but we very quickly learned that they were very difficult devices and not at all inexpensive. They were not going to offer any advantages as price was concerned, and they were actually more difficult to use in circuits" (Mueller 1975). Tunnel diodes were eclipsed by advances in integrated circuits, and by the end of the twentieth century they were no longer in widespread use.

ZENER AND AVALANCHE DIODES

Although the chief distinguishing characteristic of a diode is that it conducts electricity in only one direction, there are several types of diodes designed to conduct current "backward." The first of these is the Zener diode, named after Clarence M. Zener, who discovered the effect in the 1930s when investigating the properties of electrical insulators. Around 1950, William Shockley noticed the effect in Bell Labs' first junction semi-conductor devices, and named them Zener diodes. While a diode's p-n junction will pass electrons with little resistance in one direction, it will block the passage of electronics almost completely in the opposite direc-tion. But all semiconductor diodes have a property known as the "break-down" voltage, which is the voltage that, if exceeded, allows a flow backward through the device. Often, the diode is destroyed if the reverse voltage exceeds the breakdown level. In production models of the Zener diode, the junction is made so that the breakdown voltage is precisely set at a predetermined level, usually a voltage commonly used in circuits, such as approximately 12 volts. The Zener diode can then be used to regulate the voltage in a circuit. A Zener regulator allows the voltage in a circuit to rise up to a predetermined point, above which the diode begins to shunt off current to ground. The result is that the voltage in the circuit drops back down, until it reaches the level at which the diode again shuts off. Years af-ter Shockley named them, it was discovered that the Zener effect was not actually responsible for the Zener diode's behavior. Subsequently, Shockley diodes were renamed "avalanche" diodes. True Zener diodes were, how-ever, later designed. The main difference is the structure of the junction, which determines how breakdown occurs and which physical processes cause it. The effect is nearly the same, but at the atomic level the physics of the breakdown is different. Zener diodes are usually only used in low-voltage applications, while avalanche diodes have a wider range of useful voltage levels.

MESAS AND MASKS

A number of other manufacturing techniques were developed in the late 1950s that helped make silicon the material of choice. The early Bell Labs diffused base transistors were called "mesa" transistors, because the wafer, once diffusion was complete, was then subjected to an acid-etching process to form a raised, central plateau. To make the mesa, the center of the wafer was covered with wax, which protected it from the acid bath that ate into the exposed surface. Forming the central mesa helped tailor the transistor's operating characteristics.

Another technique, important both to the process of making diffused layers and to the selective etching of wafer surfaces, was oxide masking.

Diffusion techniques required exposing the silicon wafer to high temperatures, and this led to pitting of the wafer's surface. Bell Labs chemist Carl Frosch discovered that a layer of silicon dioxide (the rough equivalent of "rust" on silicon) on the surface protected it from pitting, while still allowing the diffusion of some types of atoms to take place.

In 1957, Bell Labs engineers coated an oxidized wafer with a photosensitive, etch-resistant polymer (later known as a "photoresist"), covered part of the wafer with an opaque mask, and exposed the surface to a strong light to activate the photoresist. Then the photoresist was chemically "developed" (like a photograph), the surface was washed to remove the unexposed photoresist on the masked areas, and the wafer was etched to create "windows" through the oxide layer into the material below. The resulting pattern was quite simple, and it suggested a way to make patterns of almost infinite complexity. But at that time, the masking technique was seen as a way to make multiple transistors on a single wafer, speeding the production process. These transistors would later be cut apart, wire leads would be bonded to them, and they would be soldered into place on circuit boards.

THE FIELD EFFECT TRANSISTOR

Also of great importance was the commercial introduction of the field effect transistor (FET). Bell Labs researchers had been working on a germanium FET even before they stumbled on the point-contact transistor, and William Shockley had theorized that an electric field (such as that created by electricity flowing through a wire) would penetrate a semiconductor and alter its electrical properties, creating, in effect, a transistor controlled by the field analogous to the grid-controlled vacuum tube. Unfortunately, it was not possible to demonstrate this until the application of the protective silicon-oxide masking technique. John Atalla and his group at Bell Labs were investigating the physical properties of the oxide-to-silicon interface when they discovered that an extremely clean, pure oxide surface stabilized or "passivated" the silicon below. This further encouraged the use of oxides in the construction of ordinary transistors, but their work also led to the revival of interest in the FET. Atalla, working with Dawon Kahng, demonstrated a new type of FET in 1960 that they called the metal oxide semiconductor transistor (MOS). In it, two separate regions of p-type silicon were diffused through "windows" in the oxide mask of a wafer. Then, using another new technique, aluminum leads were connected through the windows by vapor deposition (in effect spraying molten metal). An additional lead, called the "gate," was attached to the oxide surface in the area between the two p-type regions. A voltage applied to the gate influenced the conductivity of the narrow region between the two islands of p-type silicon, acting analogously to the base in a conventional transistor.

EPITAXIAL CONSTRUCTION

Despite the important advantages of silicon planar transistors, however, their widespread use also faced a number of important problems, including the fact that they could handle only relatively low frequencies. Much of the reason for this was that high-frequency operation was partly determined by the thickness of the collector region. At a certain point, making it thinner tended to result in more wafers being broken during processing. In 1960, however, Bell Labs provided a partial solution to this problem with the development of the epitaxial method of manufacturing silicon transistors. The epitaxial manufacturing process involved depositing a thin layer of crystalline silicon semiconductor material onto a thicker crystalline silicon substrate. The two layers were doped differently to give them different electrical properties, so the thick substrate did not interfere with the transistor's operation. The word "epitaxial" refers to the fact that the crystal structures of the substrate and the deposited layer have the same orientation and lattice structure. Device-quality transistor components could be formed in the epitaxial silicon layer without interfering with the substrate, and as a result the substrate contributed mechanical strength to the transistor without also producing the troublesome characteristics normally associated with a thick base. This process thus allowed the production of silicon transistors suitable for high frequencies and high power.

FAIRCHILD AND THE PLANAR TRANSISTOR

Fairchild Semiconductor's Jean Hoerni devised, somewhat independently of Bell Labs, an important elaboration of the oxide-masking and photolithography techniques to create the planar transistor. Hoerni and his colleagues used multiple stages of photolithographic masking, diffusion, and etching to create more complex transistor structures by 1958–1959. The

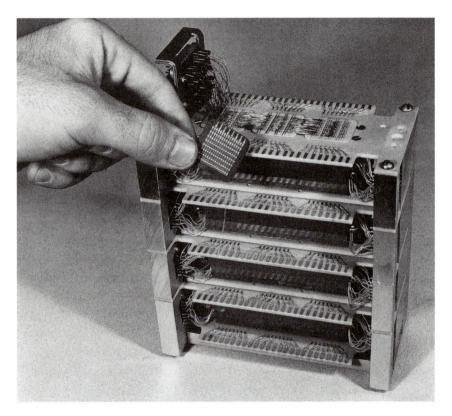

In the foreground of this 1964 Bell Labs photo is a hand holding a 256-bit "ferrite sheet," one variation of the magnetic core memory. In the background is a 196-kilobit "store" made from these sheets. A promising technology for computers and telephone switching applications, the ferrite sheet and other magnetic memory devices would remain more cost-effective than semiconductor memories through the early 1970s. Courtesy Lucent Technologies Inc.

planar transistor was probably the most important advance in semiconductor manufacturing during the 1950s, since it allowed for the inexpensive batch production of large numbers of reliable transistors. As microelectronics pioneer (turned historian) G. W. A. Dummer later noted, "The planar process is the key to the whole semiconductor work" (quoted in Braun and McDonald 1978, 74). Additionally, it is significant that the planar technique was only useful for producing silicon transistors because the metal films used in the process were not compatible with germanium. As a result, the industry as a whole rapidly turned toward the use of silicon. The importance of this shift cannot be overstated: silicon transistors were better suited to the market, and the planar technique made the batch production of reliable silicon transistors possible. As a result, within a few short years silicon transistors began to flood the market. Although germanium semiconductor devices would remain important, silicon was the future of the industry.

THE GROWTH OF THE INDUSTRY

During the period from the second half of the 1950s to the early 1960s, semiconductor production grew into a major world industry. Design improvements culminating in the silicon planar transistor led to a dramatic increase in the number of components produced and a dramatic decrease in price. The rate of growth was simply staggering. Between 1957 and 1965, the number of companies producing transistors in the United States jumped from twenty-two to forty-three, producing more than 6,000 distinct types of transistors. During the same period, the number of germanium transistors manufactured increased from about 28 million to 334 million. The number of silicon transistors made during the same period, however, jumped from around 1 million to an astounding 275 million. Along with this growth came a remarkable drop in price: the average price of a germanium transistor decreased from about $1.85 to about 50 cents, while the price of a silicon transistor plummeted from about $17.80 to about 85 cents. In 1960, an article in *Business Week* called the semiconductor industry "the world's fastest growing big business" (Smits 1985, 71).

New systems and applications that made use of the transistor's unique capabilities were appearing so quickly that it becomes impossible even to mention them all. The military remained the primary consumers of semiconductors throughout the decade, purchasing them for aircraft, missiles, communications systems, and other applications. In 1955 the armed forces of the United States purchased about 35 percent of the total dollar value of the semiconductor industry. This grew to almost 50 percent by 1960, before declining to just 24 percent in the mid-1960s as the consumer market became increasingly important.

In addition to the simple facts of plentitude and low cost stood the basic advantages of solid-state semiconductor devices over older forms of electronic technology such as tubes: durability, low power requirements, small size, and so on. This did not mean that semiconductor technology completely replaced tube technology, of course. As one observer would later recall, "For that decade or so, from '53 to '63, we had no choice but to go with vacuum tubes [in certain applications] because they did a better job, and up until that time they were cheaper. You could get a perfectly good vacuum tube for about 75 cents" (Braun and McDonald 1978, 50). In some circuits, notably the high-voltage circuits in television sets, transistors could not compete with tubes until special high-voltage designs were introduced in the late 1950s. Even after this time, well into the 1960s, many televisions were known as "hybrid" devices with both tubes and transistors (this referred only to the internal circuits; nearly all commercial TVs used cathode ray tube displays through the end of the twentieth century). Despite this, advances in semiconductor technology and the rapid proliferation of transistors made possible in part by planar technology eventually reduced the utility of tubes to niche applications. As semiconductors became increasingly important, resources devoted to the research and development of tube technology gradually declined and increasingly became limited to niche areas such as imaging and display applications. Still, tube production continued in order to support older applications designed before the transition to semiconductors. As late as 1974, for example, Western Electric manufactured approximately 1,200,000 tubes for older equipment.

Another important factor that led to the general replacement of tube technology by semiconductor technology was the simple fact of size. Reductions in the size, weight, and power consumption of semiconductor devices were extremely important factors in leading to the transition from tubes to transistors since they dramatically increased both the utility and reliability of electronic technology. Semiconductor devices could be produced at a level of miniaturization that was simply impossible for tube technology to compete with, and as transistors shrank in size they increasingly replaced the more bulky tubes previously used. As one researcher later noted, "People were enamored with miniaturization. That was the key word, how tiny it could be" (Braun and McDonald 1978, 93). This emphasis on size reduction, plus the fact that it was possible to create a multitude of transistors and circuit elements simultaneously on a single silicon wafer, eventually led a number of researchers to propose the possibility of placing a complete circuit on a single semiconductor chip, an idea finally realized with the invention of the integrated circuit.

THE MASER

Transistors were not the only important electronic devices to be developed during the 1950s. Another major category of device is the maser and its more famous successor, the laser. Masers and lasers are "quantum" devices that can be used in applications as diverse as communications, weapons systems, and surgery. Unlike the transistor and its successor the integrated circuit, both masers and lasers were used in only limited applications for many years, despite intense interest by researchers and rapid technological development. All those years of work paid off, however, and lasers have found their way into the daily lives of millions of people worldwide.

The development of the maser and laser grew out of ongoing experiments with Klystrons, Magnetrons, and other tubes developed for radar. Following the war, radar continued to drive millimeter-wave tube research, as did the relatively new field of molecular spectroscopy, which used millimeter waves to study the structure of molecules. The military was extremely interested in the continued development of millimeter systems and funded a number of research projects in the years following the war, one of which was the Radiation Laboratory at Columbia University (not to be confused with the more famous "Rad Lab" at MIT). One of the researchers there was Charles H. Townes, who came to the university in 1948 after leaving Bell Laboratories.

In 1951, while sitting on a bench in a park and mulling over the problems associated with the conventional devices used to produce microwaves, Townes claimed to have grasped a critical idea: the phenomenon of stimulated emission, in which molecules excited by an external energy source and induced to emit energy could be used to generate microwave beams (and hence the name of the maser, which stands for "microwave amplification by the stimulated emission of radiation").

The phenomenon of stimulated emission is based on the fact that while heat, light, or other energy can raise the energy level of an atom, its electrons are locked into discrete energy levels called bands. When they jump from one energy band to another, they absorb energy if they move from a lower state to a higher one, or they emit energy if they move to a lower energy state. At any given time, a small minority of molecules in a sample of any material will be at a high energy state, while the majority will be at a lower energy state.

Townes constructed an elaborate vacuum chamber in which he generated a beam of ammonia molecules. As the beam traveled through space, powerful electromagnetic "focusers" pulled low-energy molecules out of the beam but allowed the minority of high-energy molecules to pass. These molecules passed into a resonant chamber. Microwave energy directed at the chamber from an external oscillator struck the excited molecules, which released energy and set off a chain reaction in the chamber. If a relatively low level of microwave energy was directed at the chamber, it would simply be amplified, but if enough energy was applied the device would itself become an oscillator, generating microwaves at high power and at a precise frequency. Further, the packets of energy released in the chamber were all "in step," with each other, resulting in a beam where all the energy was exactly the same frequency. An ordinary vacuum tube oscillator, on the other hand, typically produced waves at a range of frequencies that varied slightly, or the unwanted higher and lower frequencies had to be electronically suppressed.

Townes, working at Columbia University, was joined by a second researcher named James P. Gordon. They demonstrated the first working maser in 1953, and in 1954 they announced their discovery in a famous article published in *Physics Review*.

The military stepped in almost immediately, funding maser research programs at various institutions so generously that some researchers jokingly referred to the maser acronym as "Means of Acquiring Support for Expensive Research." Military support for maser research, however, went well beyond the financial. In 1956, for example, the U.S. Army Signal Corps Engineering Laboratory helped encourage research into ammonia masers in a variety of ways, including organizing conferences, acting as a clearinghouse for information, and helping research teams to purchase equipment.

The ammonia maser suffered from a number of problems, including a very narrow amplification bandwidth and limited tunability. In order to get around these problems, a variety of researchers began to investigate maser action in other materials, including "solid-state" materials (as opposed to gasses). Researchers interested in developing solid-state masers besides Townes included Nicolaas Bloembergen at Harvard, Malcolm Strandberg at MIT, and numerous others. Bloembergen in 1956 revealed the three-level maser, which allowed a greater range of materials to be used and could be continuously operated, rather than pulsed as in the earlier masers.

Other researchers were exploring similar ideas at the same time as Townes. The maser was apparently conceived independently by Joseph Weber at the University of Maryland, and also by Alexander M. Prokhorov and Nikolai G. Basov at the Lebedev Physics Institute in Moscow. Similar research was also being undertaken at the time by James Gordon at Bell Labs. Over the next several years, various researchers demonstrated maser action in all sorts of materials, although the most significant of these was probably that of Chihiro Kikuchi at the Infrared Radiation Laboratory of the University of Michigan. On December 20, 1957, Kikuchi demonstrated a three-state maser using a bar of ruby, a material that was readily available, durable, and easy to adjust for tuning purposes. The ruby maser, unlike the earlier designs, seemed readily adaptable to practical applications. Solid-state masers operating at microwave frequencies of 1 to 21 millimeters were being widely deployed for space communication, radio astronomy, and military communication by the end of the decade.

3

Microchips and Lasers

◆

The decade of the 1960s began with a crisis that demonstrated how intertwined technology and international relations had become since the end of World War II. In 1960 the Soviets shot down an American U2 "spy plane," a marvel of engineering carrying advanced surveillance gear. The attack came from another form of high technology, a sophisticated missile system. When the Soviets justifiably accused the United States of spying and demanded an apology, President Eisenhower refused, leading to heightened tensions between the two countries. Later in the decade, relations between the United States and the Soviets began to improve as both sides toned down their aggressive rhetoric and pledged to avoid nuclear war. A technological symbol of this new cooperative spirit was the installation of the "hot line" telephone between Moscow and Washington. By 1969, the United States and the Soviet Union found the political will to conduct the Strategic Arms Limitation Talks (SALT). Meanwhile, although military funding waxed and waned, research, development, and procurement continued at a rapid pace. By this time nearly every advanced military system, from command and control to ships, tanks, aircraft, and missiles, depended heavily on electronics and, increasingly, computers.

ELECTRONICS IN THE COLD WAR

The U.S. government continued to fund electronics research and development for military purposes throughout the decade, but Cold War tensions led the government to support the field in other ways as well. One of the most important of these was through its commitment to the so-called space race. After launching its first successful satellite in 1957, the Soviet Union scored again in 1961 when astronaut Yuri Gagarin became the first man in space. Embarrassed by this Soviet upstaging, President Kennedy called for an American moon mission that same year. Remembered mainly for its dramatic rocket launches and landings, technologically the exploration and use of outer space were also heavily dependent on electronics. By 1962, John Glenn had become the first man to orbit the Earth, and the Telstar satellite began beaming telecommunications signals across the Atlantic. Both were milestones in electrical engineering. For the next several years, unmanned missions garnered most of the public attention, such as the 1964 Mariner IV mission. In many ways, unmanned craft were even more dependent upon advanced electronics. Carrying a television camera, scientific instruments, and communication equipment, the small Mariner IV probe provided the first close-up photos of Mars. The Soviet Venus III probe, loaded with electronic test instruments and telemetry, made an intentional crash landing on Venus in 1966 to become the first manmade object to touch another planet. The Soviets were the first to soft-land an unmanned craft on the Moon, but by the end of the decade, in November 1969, the United States successfully fulfilled Kennedy's request and put a man on its surface. Such high-stakes public projects made the governmental influence in engineering more important than ever.

Charles H. Townes as Dr. Strangelove

Charles Townes invented the maser and published the first scientific paper outlining the theory of the laser.

I think one of the strengths of American science and technology that came about after World War II is that people had been mixed together and got to know each other and each other's fields. When they went back to the universities and into industry, they had that kind of background. I think that interaction was very important, and the interaction with government.

It was cut off by the Vietnam War. The Vietnam War made university people try to shy away from business and industry and shy away from government. I was here at Berkeley during much of it, and, oh, people would really jump on me for having anything to do with industry or the military. For example, when I went on the board of General Motors, I knew there was going to be a lot of criticism. I contacted the president of the university and told him that the chairman of the board of General Motors asked me to form this advisory committee. . . . I felt it was a sensible thing to do. But I recognized that it might be criticized by the university, and I asked what he thought. Did he think I ought to do it? He said, "Well, I think on balance you ought to do it." So I did it. But sure enough I got jumped on in the student newspaper, the idea that I would have anything to do with a big commercial company like that. That was sinful. Nowadays it's welcomed. [Chuckling] People want it because they think maybe GM can give them some money. But then it was very bad, and so was my connection with government. I was compared with Dr. Strangelove in various ways.

Source: Charles Hard Townes, an oral history conducted September 14–15, 1992, by Frederik Nebeker, IEEE History Center, Rutgers University, New Brunswick, New Jersey.

Electrical engineers clamored for greater recognition for their role in such triumphs of "big science." To the public, however, these types of devices and systems remained obscure and hard to understand, and the public stature of the field was only temporarily bolstered by engineers' involvement in the space race. More evident to the public was the growing number of consumer products that made use of new electron devices such as the transistor. By the 1960s, these were rapidly replacing vacuum tubes in home and car radios, tape recorders, and stereophonic sound systems. However, one of the most exciting of these products was color television, which still relied heavily on tubes. Color TV had languished since the early 1950s but became big in the late 1960s as improvements in tube manufacturing dramatically lowered costs. However, most of the other important technical innovations remained beyond the view of the public during the 1960s.

This was the decade that produced many of the electron device innovations that would become household words in later years but that, for the moment, remained obscure. Most notable in this regard was the integrated circuit, the device that would later become the basis of the microprocessor and the memory chip. As with so many other Cold War innovations, the integrated circuit (IC) saw its first significant application in weapons systems such as the Minuteman II missiles of 1962, and contractors (notably North American Aviation) received $24 million in IC-related defense contracts over the next three years. About 500 Minuteman II missiles were deployed through 1969.

PEAKS AND DECLINES

In the United States, electron device research was reaching heights of creativity that would not be matched again. The Bell Telephone Laboratories of AT&T; the RCA Research Laboratories in Princeton, New Jersey; the Westinghouse device research facilities in Pittsburgh; and other centers of electron device research and production were all at the peak of their art, cranking out a rapid stream of innovative devices, some of which are still in production today. Building on the successes of the early 1950s, by the early 1960s the transistor was widely employed in all sorts of consumer electronic devices, and over the course of the 1960s would become the basis of increasingly complex integrated circuits. New types of transistors, offering faster switching speeds, higher current-carrying capacity, higher efficiency, or wider bandwidth operation poured out of the labs, although many innovations would require years of additional research to become practical.

Despite this success, the late 1960s and early 1970s also marked the beginning of a general decline in the ability of U.S. corporations to transfer knowledge from the laboratory to their own manufacturing facilities and to compete in the developing global electronics market. This was particularly true in the consumer electronics industry. One by one, U.S. electronics manufacturers (with the notable exceptions of color television and automobile radio makers) gradually began to close their U.S. production plants, or began to assemble equipment manufactured from parts made in places like Japan and Singapore. Some products, such as consumer radio receivers and tape recorders, were almost completely taken by non-U.S. manufacturers by the mid-1960s. This also indicated the general direction that the industry as a whole would take in later years, with the United States losing its dominant position in the semiconductor industry and instead increasingly focusing on the assembly of parts made in other locations. In many cases, electronics companies welcomed this shift, moving factories to low-wage countries as a way to boost profits. American consumers, who saw prices fall for popular products, set aside their nationalistic objections. In other cases, U.S. firms launched facilities in other countries, chiefly in Europe, as a way to enter semiconductor markets there on better economic terms. Know-how in the semiconductor field spread rapidly, benefiting U.S. companies in the short term, but in effect jump-starting their competitors in other lands.

VACUUM TUBES IN THE 1960S

The use of vacuum tubes for reception and amplification purposes was gradually disappearing in the 1960s, and the development of most new tube types ceased. Older types remained in production, however, and supported a lucrative replacement market for many years. There was continued development in tubes for microwave communications and radar, both of which were by now essential in military operations and in systems such as civilian air traffic control. The Anti-International Ballistic Missile defense system and Nike-Zuse missile systems of the late 1950s, the Ballistic Missile Early Warning System and the North American Aerospace Defense Command (NORAD) projects of the same period, and the space vehicle tracking system associated with project Mercury all relied on vacuum tube transmitters.

One of the interesting footnotes in the vacuum tube story in the 1960s is the development of the first commercial microwave ovens. The Litton Corporation, an American company associated mainly with military technologies, designed a low-cost version of the Magnetron tube for use in a commercial oven, and contracted with the Japanese firm Kobe Kogyo for its manufacture. While Litton had little commercial success, the Japanese firm Hayakawa picked up the idea and offered its first microwave oven in 1962.

More generally, once the transistor appeared, new types of tubes were developed to directly compete with them. In the late 1950s and 1960s, tubes shrank in size and were sometimes touted as challengers to solid-state devices. RCA introduced its Nuvistor, for example, in 1961. This was a ceramic tube, about the size of a transistor, that was designed for use in radar technology. The Nuvistor found a broad market in consumer electronics, where it was used in television tuners and other devices. Tube designers also developed their own version of the integrated circuit. General Electric introduced the Compactron in 1961, a tube device that was in essence several tubes built inside one glass envelope. These multi-unit tubes also proved very popular in consumer electronics, and fended off semiconductors in television sets through the early 1970s.

However, while tube designers looked for ways to compete with semiconductor technology, designers of solid-state components also saw vacuum tube technology as territory to be conquered, and solid-state technology increasingly challenged tube technology in even the most specialized niche applications. As the transistor matured in the 1960s, certain kinds of solid-state devices were developed as direct replacements for tubes. The so-called overlay transistor, invented by RCA in 1964, for example, was intended to be a direct replacement for vacuum tube output devices used in UHF military communications equipment. One early overlay transistor, the 2N3375, produced a remarkable 10 watts of output power at 100 MHz or 4 watts at 400 MHz, making it comparable to a small transmitting tube. Microwave communication was another area where tube designers were confident that the transistor would prove inadequate, given the limited bandwidth of the early devices. In 1963, however, J. B. Gunn demonstrated oscillations in compound semiconductors under an applied electric field, and the first commercial Gunn diodes began appearing around 1966. Another important semiconductor advance that infringed on tube territory was the impact avalanche and transit time (IMPAAT) diode, developed in 1964 at Bell Laboratories by R. L. Johnston and B. C. Deloach. As can be seen from the development of these two devices, the continued importance of tube technology did not go unchallenged by innovations in solid-state devices. The field of imaging and display devices reflects these broader trends: although it remained one of the most important specialized applications for tubes and helped keep the vacuum tube industry afloat, imaging and display devices also became an important area of development for solid-state technology during this time.

THE INVENTION OF THE INTEGRATED CIRCUIT

Perhaps the most important new electron device of the 1960s was the integrated circuit, invented in 1959 and commercialized by 1960. An integrated circuit is a device in which some or all of the various discrete components that make up a circuit, such as transistors, diodes, resistors, conductors, inductors, and capacitors, are manufactured simultaneously on a single semiconductor wafer. In earlier forms of circuit design, "active" devices such as transistors were fitted with wire leads and connected by wires to "passive" resistors, capacitors, and other components, but in the IC both active and passive devices are etched into the silicon. Silicon even becomes a replacement for the wires that once linked device to device.

The integrated circuit was the direct outgrowth of transistor manufacturing techniques invented in the 1950s. In particular, the planar transistor incorporated a number of extremely important features that would be applied in IC fabrication. At its core, the integrated circuit was a manufacturing innovation rather than a scientific breakthrough. Practicality in manufacture was particularly important, as Jack Kilby, one of its inventors, later remembered: "In contrast to the invention of the transistor, this was an invention with relatively few scientific implications. . . . Certainly in those years, by and large, you could say that it contributed very little to scientific thought" (Braun and McDonald 1978, 90).

Instead, the integrated circuit should be seen as a practical solution to the interrelated issues of commercial production, the military demand for more complex systems, and the continuing emphasis on the miniaturization of circuits. There was also an imbalance between the abilities of circuit designers and the abilities of electronics manufacturing facilities that would favor the integrated circuit. Computers and other complex systems required that thousands of transistors, diodes, and other components be laboriously soldered together by hand. As the complexity of such systems grew and the number of components needed for each application likewise increased, the time needed to manually connect all the components grew to be unreasonable. As historians Braun and Macdonald have argued, by the late 1950s the total number of man-hours needed by the industry as a whole outstripped the total hours available from its workforce, and the industry faced a serious bottleneck (1978). At the same time, reliability became a serious problem as well. Since each component had to be manually connected, the possibility of human error increased along with the number of components used. Furthermore, the greater the number of discrete components that needed to be manually connected, the bulkier the final system, leading to corresponding limits on practicality in many systems that required a compact size. By 1962, for example, some computers had as many as 200,000 individual components. The use of circuit boards, where all the components were fitted onto a thin board and automatically soldered, helped somewhat. But the possibility that a human error would occur at some point during the process of manufacturing these huge devices was, not surprisingly, quite high.

The demand for reliable, miniaturized systems even smaller than those made with the diminutive transistor also grew out of the military's early successes with guided missiles and the corresponding desire to create even more accurate guidance systems for them. The relative share of money spent on technology for missile systems as a percentage of total military procurements had grown from 5.4 percent to 27.4 percent between 1955 and 1960, while at the same time, the role of the military as a consumer of semiconductor products increased to its highest level, with the military buying close to 50 percent of the total dollar value of the industry's output in 1960. In addition to the stimulus provided by military contracts, the government helped guide electron device research more directly through the sponsorship of conferences, the support of academic institutions, and the sponsorship (sometimes indirect) of official boards of experts such as the Advisory Group on Electron Tubes and the later Advisory Group on Electron Devices.

INTEGRATION BEFORE THE IC

Because so many different researchers were working on the problems of miniaturization and reliability, it is not surprising that there were a variety of proposed solutions. The printed circuit board, for example, allowed a certain degree of miniaturization because a lithographic process was used to etch electrical connections on a copper-clad, phenolic plastic sheet. Holes were drilled in the board to allow the insertion of device leads, then the leads were attached to the conducting foil by soldering. This also had the important benefit of allowing the automation of the assembly process. By using transistors and miniaturizing other components, very small circuit boards could be designed to hold circuits on, say, a 1-inch square that previously required 100 square inches or more. The U.S. Navy, for example, had sponsored what it called Tinkertoy technology beginning in 1950. This was a method of assembling transistors and other components on printed circuit boards and stacking them into very compact modules that could be inserted into "sockets" much like those used for vacuum tubes. This involved the use of the smallest practical discrete components and the development of automatic machinery to assemble and wire together the circuits.

The development of mask-and-etch techniques for making transistors suggested the possibility that an entire circuit could be built directly onto a single wafer of semiconductor material. This possibility was suggested as early as 1952, when Geoffrey William Arnold Dummer of the Royal Radar Establishment in England argued that layers of insulating, conducting, rectifying, and amplifying materials could be stacked to form a circuit, interconnections being made by cutting away appropriate sections of the layers. Dummer, however, failed to construct a working device. Bell Laboratories in the mid-1950s constructed a four-in-one transistor from a single silicon wafer through photolithography and diffusion processes. The four transistors were directly connected to each other to form a special circuit used by one of AT&T's pieces of telephone equipment. While arguably the first multitransistor circuit to be "integrated" on a single silicon wafer, it was somewhat different than the type of integrated circuit that would eventually come to dominate the field.

In 1958 the young engineer Jack Kilby of Texas Instruments began thinking, as other engineers clearly had, about how to construct an entire circuit in silicon. What made his idea different was that he would not only build multiple transistors on a single wafer, but also use silicon to form the conductors, capacitors, and resistors for the circuit. As it turns out, silicon itself can be used as a resistor, in place of the carbon resistors usually used in transistor circuits. A single p-n junction can also function as a capacitor, replacing the discrete foil, electrolytic, and other capacitors usually employed. Using a semiconductor as a resistor was seen as a waste of valuable "real estate" on the wafer, but it allowed nearly the entire circuit to be fabricated at once. To demonstrate the idea, he built a simple, standard type of circuit called a phase-shift oscillator in a sample of germanium. Texas Instruments filed a patent for "solid circuits" in February 1959.

As in so many cases in the history of technology, others were working along similar lines. At about the time of the filing of the Texas Instruments patent, Robert Noyce at Fairchild conceived a way to make a circuit consisting of multiple semiconductor devices, all connected by aluminum "wiring" that was deposited on the silicon through the same mask-and-etch techniques used to make the transistors themselves. Fairchild patented the device, and even though the idea was similar to Kilby's earlier idea, Noyce's "unitary circuit" was the first to be awarded a patent in 1961. Predictably, a dispute soon erupted between Texas Instruments and Fairchild over who had invented the idea of an integrated circuit. The two companies quickly became embroiled in a patent dispute lasting for many years. Yet by 1962 both companies were producing small quantities of integrated circuits for military, space, and commercial applications. Fairchild's first "micrologic" ICs, sold in 1962, cost $120 each and incorporated only a few circuit components. By 1965, twenty-five U.S. companies, led by Fairchild, were producing integrated circuits.

A number of significant problems confronted the industry in making commercially viable integrated circuits. One was production yield, which remained troublesome for the first half of the 1960s. Wafers used to produce integrated circuits needed to be larger than ones used to manufacture discrete components, and as a result they were more likely to be chipped or broken, or to suffer from contamination during the manufacturing process. More importantly, although making use of the same production techniques, the manufacture of integrated circuits required more steps than that of planar transistors because there were more layers of transistors, diodes, passive components, and interconnections. As a result, yield was substantially

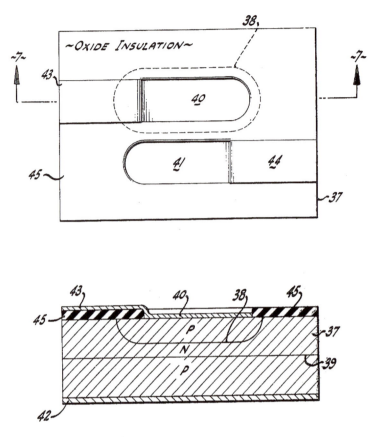

Robert Noyce's 1959 patent for an integrated circuit, which he called simply a "device and lead structure." On a substrate of p-type silicon, photolithographic techniques are used with oxide masking and diffusion to create localized layers of n-type, p-type, and insulating regions. Diodes, transistors, and other devices created in this way are interconnected by wire leads deposited on the surface. U.S. Patent 2981877.

lower due to the accumulation of potential manufacturing errors at each step. Early production methods, in fact, could sometimes only manage a yield of 25 percent (or sometimes worse). In 1963, however, Motorola discovered that reductions in wafer size led to a disproportionate increase in production yield due to the fact that defects produced through the manufacturing process were not randomly distributed on the wafer; reducing the size of the wafer to 25 percent of its original size, for example, improved production yields by more than a factor of four.

While wafer sizes would soon begin to grow, this discovery helped motivate the search for ways to make integrated circuit components smaller

Robert Noyce, Gordon Moore, and Andrew Grove, founders of Intel Corporation. Courtesy of Intel Corporation.

and smaller, resulting not only in better production yields but also in faster and more powerful devices as more and more components were squeezed onto a single wafer. As in the case of transistor production, for example, silicon wafers were prepared for chemical etching using photosensitive patterns. Pattern generators and step-and-repeat cameras, both of which were introduced in the early 1960s, helped reduce the size of integrated circuits

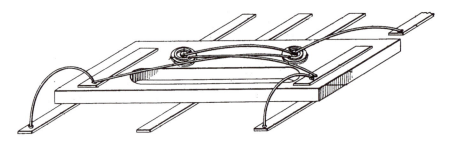

Jack Kilby's patent for "miniaturized electronic circuits" in 1959 showed a "multivibrator" rendered in a single slice of semiconductor. It consisted of two transistors (the circular structures) formed by planar techniques and connected with wire leads to each other and an external circuit. In principle much like Noyce's design, Kilby had already demonstrated a working integrated circuit in the laboratory. U.S. Patent 3138743.

Looking much less elegant than the patent drawings is Jack Kilby's first integrated circuit of 1959. Courtesy of Texas Instruments.

through the precision made possible by automation. Ultraviolet lithography was also substituted for ordinary photolithography, which helped shrink the size of integrated circuits since ultraviolet light has a shorter wavelength than visible light, thus allowing the production of smaller components. Ion implantation also helped reduce the size of circuit elements, a process first developed by Russell Ohl and William Shockley at Bell in the 1950s.

Ion implantation used an ion beam, rather than vapor diffusion, to produce n- or p-type junctions, and as a result allowed highly accurate placement of circuit components. These were all important techniques in improving both the production yield and the speed of integrated circuits by making the wafers smaller and increasing the number of components on each chip.

Another significant problem facing the semiconductor industry was the laborious process of manually attaching leads from the semiconductor wafer to the outside world (such as the leads for the circuit's power supply, and data input and output). In the early 1960s, Martin P. Lepselter of Bell Labs addressed this problem by developing a technique to manufacture integrated circuits with gold leads that were already attached. These gold

Jack Kilby, photographed at about the time of the invention of the integrated circuit. Courtesy of Texas Instruments.

beam leads were formed during the manufacture of the chip rather than attached afterward, and the free ends could all be attached to another device component at the same time rather than one by one, thus saving considerable amounts of labor. This innovation also helped address the problem of costly hermetic sealing of the circuit inside a small canister. In order to avoid contamination, integrated circuits needed to be contained in expensive enclosures. However, in 1966 J. V. Dalton showed that silicon nitride acted as a barrier to the migration of sodium ions, the particles that were a main culprit in contamination and device instability. Several Western Electric engineers soon demonstrated that a thin film of silicon nitride could be used as a shield on planar n-p-n transistors, with chemical techniques being used to open contact windows through the film. This innovation was rapidly combined with gold beam leads to form beam lead sealed-junction (BLSJ) technology, which was later used in a wide variety of integrated circuits. BLSJ devices were the first high-reliability integrated circuits that did not require hermetically sealed encasing. One direct result of the development of BLSJ technology was a family of circuits known as transistor-transistor logic (TTL) circuits, which combined BLSJ technology with standard buried collector (SBC) technology.

ANALOG AND DIGITAL SPLIT

Soon after its introduction, integrated circuit design became differentiated into analog circuits and digital circuits. Many engineers considered analog integrated circuits—that is, ones that amplified or otherwise manipulated electrical signals such as radio waves—the less promising field, since they usually demanded large capacitors, large inductors, and other components that could not easily be fabricated on the silicon chip itself. Despite this, analog integrated circuit design surged forward during the 1960s. A leader in this area was Fairchild's Robert Widlar, who developed the first practical operational amplifier (or "op-amp") in IC form. The op-amp was, ironically, a basic building block of the analog computer, a form of device now almost completely superceded by the more familiar digital computer. In an analog computer, the op-amp performed mathematical functions such as adding or subtracting by comparing a reference voltage to an input voltage. As G. W. A. Dummer later put it, Widlar's development of the famous "μA 702" and "μA 709" op-amps was

> a revolution of sorts. Rather than translate a discrete design into monolithic form, the standard approach, Widlar played the linear microcircuit game by

a different set of rules; use transistors and diodes—even matched transistors and diodes—with impunity, but use resistors and capacitors—particularly those of large value—only where necessary. Even where use of a big resistor seemed inevitable, Widlar put a dc-biased transistor in its place. He exploited the monolith's natural ability to produce matched resistors and only assumed loose absolute values. (quoted in Dummer 1997, 188)

Widlar's μA 702 and μA 709 integrated op-amps were widely adopted by the industry. The μA 709 is still produced today, since it can be used in devices such as audio equipment as a sensitive, low-noise preamplifier. Other firms, such as Sprague Electric and Hitachi, developed analog integrated circuits that exploited niche markets based on consumer applications. Sprague, for example, introduced linear integrated circuits for use in Delco automotive radios, Zenith televisions, and Polaroid cameras in the late 1960s. These special-purpose analog chips constituted perhaps the first breakthrough to the consumer market, and analog chips are still a major industry, but they would soon be overshadowed by developments in the computing field.

As technology continued to improve, the semiconductor market as a whole increasingly focused on digital rather than analog integrated circuits. This was related, in part, to the growing importance of computers in business and military operations and to the importance of calculators and other such applications as commercial products. More importantly, it represented a broad shift toward the manipulation of digital data as the basis of the semiconductor industry. In other words, the control and organization of digital data, rather than of various kinds of analog signals, became the major focus of the semiconductor industry. This reflected both a split between computer-related technology and the rest of the field and the growing dominance of this technology within the industry as a whole.

During the 1960s, the digital branch of integrated circuit design was focused on logic circuits and, toward the end of the decade, memory devices for computers. Logic circuits are the building blocks of computers, and are used to make "decisions" based on information they receive in the form of electrical pulses. These pulses, representing binary "ones" and "zeroes," are subject to logical "operations" in circuits called gates. The gates are circuits designed to provide a specified output whenever they are supplied with a particular input. These inputs and outputs, in the form of electric pulses, need to be detected, amplified, and moved around in the chip, and this is accomplished using transistors as amplifiers and/or switches. Computers and related systems had used vacuum tubes in the early 1950s and had begun to use transistors as early as 1958. From the beginning, integrated circuits were employed in logic circuits, and circuit designers had to employ new ways to build circuits from silicon that could function effectively in logic applications. The early and mid-1960s saw a group of integrated circuit manufacturers battling with each other to win support for their unique logic circuit designs. Each had its advantages and disadvantages in terms of cost, difficulties in manufacturing, and performance in computer circuits.

RTL, DTL, AND TTL

In later years, these distinct types of logic circuits would be known as logic "families." Fairchild in 1961 introduced the first, resistor-transistor logic or RTL. This circuit performed logical operations using two or more transistors connected together by silicon resistors. Because these circuits consumed a relatively large amount of power, others developed diode-transistor logic (DTL), in which the basic logic functions were performed by diodes (which are "passive" devices that do not consumer power). Then the output, which was weakened in the process, was sent to a transistor to be amplified. However, because of the way integrated circuits were manufactured, it made more sense to use transistors in place of the diodes at the input stages of the circuit. In 1963 Sylvania offered the first transistor-transistor logic (TTL) circuit, under the brand name SUHL (for Sylvania universal high-level logic). Here, a special transistor with multiple emitters took the place of the diodes in the input stage. An even more important advantage of TTL over DTL was its high speed; it could perform more logic operations per second than its competitors. Sylvania's version of TTL is not as recognized as that of Texas Instruments, which would offer its version of TTL in 1965, followed by Fairchild the next year.

MOS LOGIC

Some time later, the junction or "bipolar" transistor logic IC was joined by the MOS logic IC. It took a decade from the invention of the MOS transistor to the commercialization of standard types of MOS-based logic ICs. In 1962 RCA engineers Steven R. Hofstein and Frederic P. Heiman succeeded in building the first integrated circuit based on this technology. Using the newly developed planar process, they built a multipurpose logic circuit that had sixteen MOS field effect transistors (MOSFETs) on a

0.25-inch square chip of silicon. This type of MOS integrated circuit consumed less power than bipolar integrated circuits, and was theoretically faster and cheaper to build. In practice, however, the production of MOS integrated circuits was plagued with problems, and soon companies such as RCA had temporarily abandoned the technology in favor of continued investment in bipolar integrated circuits. While MOS logic would become much more important in later years, through the 1960s it remained on the horizon while TTL chips dominated the market.

MEMORY

Toward the end of the 1960s, researchers began to develop semiconductor memory chips in addition to logic chips. Electronic memory had been a goal of computer designers since the late 1940s, but in the 1940s and 1950s engineers had to make do with magnetic drums, mercury delay lines, cathode ray tubes, magnetic disks or tapes, and, later, magnetic cores in order to store data. That began to change when, in the early 1960s, Texas Instruments built the first integrated semiconductor memory circuit. However, since only six transistors could be fit on a single chip at the time, the device was only able to store a single bit of data. This was clearly not going to be commercially feasible, so many designs stuck with magnetic cores for the time being. There were, however, IC memory devices introduced at various times in the 1960s. IBM, for example, produced a 16-bit memory chip using bipolar transistors in 1966, which was used in an IBM/360 computer delivered to NASA. The company would produce a 64-bit IC memory chip by 1968. IBM scored again in 1966 when Robert Dennard designed a circuit to store each bit of information as the level of charge on a capacitor, controlled by a single FET transistor. The single-transistor "cell," first used in this IBM dynamic random-access memory (DRAM) chip, became a standard type of memory design.

SHRINKAGE AND GROWTH

The crucial development that allowed the practical manufacture of semi-conductor memories was MOS technology, which allowed an extremely large number of circuit devices to be placed on a single chip for the first time. MOS technology returned to the fore in the late 1960s due to a wide variety of technological advances at a large number of different companies.

As a result, the precise manner in which the industry began to turn toward MOS integrated circuits is difficult to discern clearly. Bell Laboratories, for example, extended the use of its BLSJ transistor technology to the manufacture of MOS transistors in the late 1960s, and in 1967 began to develop what it called dual-dielectric MOS integrated circuits. By 1969 this had resulted in a single chip containing up to 120 circuit components. A large number of other new technologies and improvements, such as the complementary metal oxide semiconductor (CMOS) chip, invented by Frank Wanlass at Fairchild in 1963 and also developed at Westinghouse, GTE, RCA, and Sylvania by 1968, appeared in rapid succession throughout the decade. By the late 1960s, MOS integrated circuits had become a commercially viable product, and in 1969 MOS IC sales in the United States had reached $30–35 million.

Although slower than bipolar integrated circuitry, MOS transistors were physically smaller than bipolar transistors, and since they used less power, all the other elements on the chip were also smaller, allowing a substantially greater number of components to be placed on the same-sized chip. The dramatic increase in circuit density made possible by the development of the MOS transistor and the subsequent development of MOS integrated circuits finally made semiconductor memory chips competitive with existing memory devices (such as core memory).

Increases in circuit density accompanied reductions in component size: just a few components had been built onto the earliest integrated circuits, but by 1971 this number had increased to about 6,000 components per chip. In the days when a few dozen circuit components on a single chip had seemed like a lot, engineers had referred to integrated circuits containing between ten and a hundred or so components as "large-scale integration." With the arrival of MOS technology and circuit density in the range of thousands and tens of thousands, they somewhat awkwardly began to refer to "very large-scale integration" (VLSI). Luckily, engineers had the sense to stop adding superlatives even as circuit density continued to increase. The power of this combination can be seen in the sheer numbers: the production of integrated circuits increased at a remarkable rate during the 1960s, jumping from 4.5 million in 1963 to 635 million in 1971. What is even more striking is the number of circuit components this represents: from 108 million in 1963 to 40,653 million (or over 40 billion) in 1971. Without the integrated circuit, the industry would simply have been unable to make practical use of many discrete components.

Gordon Moore in *Electronics* Magazine, April 1965

Gordon Moore's article in *Electronics* magazine in 1965, excerpted below, led to the famous "Moore's Law." At the time, Moore was director of research at Fairchild Semiconductor.

Integrated circuits will lead to such wonders as home computers—or at least terminals connected to a central computer—automatic controls for automobiles, and personal portable communications equipment. The electronic wristwatch needs only a display to be feasible today. . . . The complexity for minimum component costs has increased at a rate of roughly a factor of two per year. Certainly over the short term this rate can be expected to continue, if not to increase. Over the longer term, the rate of increase is a bit more uncertain, although there is no reason to believe it will not remain nearly constant for at least 10 years. That means by 1975, the number of components per integrated circuit for minimum cost will be 65,000. I believe that such a large circuit can be built on a single wafer.

FROM TUBES TO SOLID STATE IN DISPLAY AND IMAGING DEVICES

Although innovations in solid-state technology would prove to be extremely important in the field of imaging and display devices in later years, older tube technologies continued to be developed throughout the 1960s and, for the most part, dominated the field during this time. The cathode ray tube is particularly important in this regard, with CRT technology rapidly improving throughout the decade. In the late 1960s, for example, Sylvania developed the Penetron CRT for displaying high-resolution color images. Penetron CRTs had screens made up of layers of differently colored phosphors, separated by thin layers of an insulating barrier. By varying the acceleration of the electron beam, the beam would penetrate into just one of the three phosphor regions, activating the appropriately colored phosphor there. The advantage of this was that it reduced the three-phosphor cluster to a single spot and avoided the use of a shadow mask, both of which limited designers from using smaller phosphors and, therefore, limited the image resolution of the screen. From the 1960s to the 1980s, high-resolution CRTs based on the Penetron were used in aircraft and ground-based radars.

A wide variety of other color television CRTs were developed during this time as well, such as Sylvania's Zebra tube, first developed in 1962, and RCA's 25AP22, the first commercially successful rectangular color television CRT. The 25AP22 had a large and relatively flat 25-inch screen and yielded higher color contrast than earlier color television CRTs due to an aluminum-foil electron shield placed between the edge of the shadow mask and the electron gun. This prevented the inevitable stray electrons from reaching the screen and degrading picture quality. RCA superceded the 25AP22 with a series of improved models, and by 1965 all of the major U.S. manufacturers of CRTs were producing rectangular CRTs for color television. One color CRT developed during this period that deserves particular emphasis is Sony's Trinitron, first introduced in 1968. The Trinitron uses a single electron gun to produce three electron beams. All three beams pass through a common focusing lens, after which a set of electrostatic convergence electrodes brings them together into a single beam that strikes the screen. Most other color tubes had used three separate beams that had to "converge" on the correct spot on the back of the screen in order to illuminate the appropriate phosphor. The combination of the three beams into a single beam meant that the convergence adjustment and other necessary operations were considerably easier than with conventional shadow-mask tubes, and as a result picture quality was higher and manufacturing and servicing time considerably lower. The first Trinitrons were 12-inch rectangular tubes, but larger models were soon introduced. In addition to color

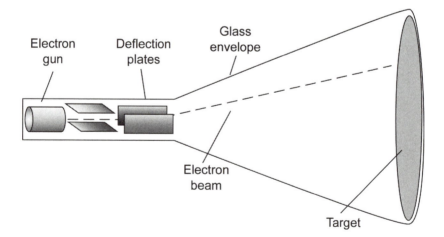

A cathode-ray tube (CRT) of the type once used in oscilloscopes. Electrons emitted from the sun are accelerated toward the phosphor-coated target. The beam is modulated by varying the electric field generated by deflection plates.

television CRTs, Trinitrons were also widely adopted for use in high-resolution computer monitors, air-traffic control applications, and aviation applications. They quickly became so popular that they began to displace U.S. manufacturers from the field of manufacturing television CRTs.

Other important CRT innovations during the 1960s included the development of ceramic CRT envelopes, the solving of the phosphor persistence problem in radar tube technology, and the development of flat-faced tubes and internal "graticules" (grid lines) for use in oscilloscopes. The flat-faced tubes and internal graticules for oscilloscopes were important because they allowed more accurate measurements to be made directly from the screen, something difficult to do using older curved-face CRT technology.

SOLID STATE DEVELOPMENTS

Despite advances in tube technology, solid-state devices began to make inroads into the imaging and display field during the 1960s. In imaging, this early solid-state work was based on simple light detection rather than the more complex business of capturing full images. Research in photodetector technologies for telecommunications purposes led to the development of the first functional photodiodes in the early part of the decade. R. P. Riesz and others at Bell Labs developed fast-acting photodiodes in 1962, which were apparently aimed at possible uses in detecting rapidly changing beams of laser light, then being considered for communication purposes. A large number of silicon and germanium photodiodes were researched and developed during the next few years.

In addition to photodiodes and phototransistors developed for these simple light-detecting purposes, work began in the 1960s toward what were essentially solid-state replacements for conventional television camera tubes. In 1967, Eugene Gordon and his colleagues at Bell Labs developed a hybrid vacuum tube/solid-state device in which the image sensor was an array of about half a million tiny, planar silicon photodiodes, built on a wafer about 0.75 inches in diameter. The elements of this array changed their electrical properties when struck by photons, becoming, in a sense, like discharged batteries. A scanning electron beam then passed over each diode in series, and the amount of current each element needed to regain a full "charge" could be measured electronically to generate the video signal. While sensitive, its operation still required a bulky glass envelope, which to a certain extent limited its practical usefulness. However, it was well suited to its intended use, which was the ill-fated Picturephone system, an early attempt at two-way video telephony.

THE LIGHT-EMITTING DIODE

1962 saw the nearly simultaneous announcement by several different re-search groups of semiconductor devices that could emit light. Researchers from RCA and Lincoln Laboratories at MIT both presented technical papers on this subject that stimulated considerable interest. Nick Holonyak at General Electric had been inspired by the RCA and Lincoln Lab presentations. Holonyak, in competition with Robert Rediker at Lincoln Labs, Marshall Nathan at IBM, and even his colleague Robert Hall at GE, rushed to develop the semiconductor laser. Hall won that race, and Holonyak's team demonstrated their own laser a little later. Along the way, though, they had discovered another type of device, a diode that emitted bright but "incoherent" light. That is, it was more like a semiconductor light bulb than a laser.

This light-emitting diode (LED), announced in 1962, was made from a different material than the gallium arsenide used in the early lasers. It was an alloy made from gallium arsenide and phosphorus (GaAsP). Although the first commercial GaAsP LEDs produced by GE retailed at $260 and therefore were not cost-competitive with ordinary lamps, GaAsP devices would become quite commercially important in later years. While Holonyak subsequently moved to the University of Illinois, parts of his work were taken up by Monsanto, which was a supplier of gallium arsenide wafers. One important area of progress was in conversion efficiency. Drawing on Holonyak's work, researchers at Monsanto Laboratories discovered that nitrogen doping greatly enhances the efficiency of red LEDs. By the end of the decade, sufficient advances in this area had been made to allow David Thomas at Bell Laboratories to produce gallium phosphide (GaP) LEDs that could emit green light. After Holonyak moved to the University of Illinois, his student M. George Craford developed a yellow LED in 1970. Another important development was S. J. Bass and P. E. Oliver's technique for growing large-area, single-crystal substrates. This was a substantial step toward making the production of LEDs commercially viable through the production of efficient junctions in large crystals. Organic light-emitting diodes (OLEDs) were also announced in the early 1960s. OLEDs were less expensive, lighter, and less brittle than LEDs made from nonorganic materials, but were also plagued by problems of low efficiency and low light intensity. As a result, their usefulness would not be realized for many years.

Although these early LEDs offered a number of desirable features compared to other display technologies then available, including reliability, durability, high contrast, and easy interoperability with integrated circuits, their usefulness in commercial applications was initially quite limited due to their cost, small size, and tendency to cause eye discomfort to viewers. As a result, it would still be a number of years before LEDs became widely used.

LIQUID CRYSTALS

Another area of development during the 1960s that would have important future implications for display technologies was in the field of liquid crystals. Liquid crystals had first been discovered back in 1889 by the Austrian botanist Friedrich Reinitzer and the physicist Otto Lehmann. They observed that some organic compounds, particularly those with rod-shaped molecular structures, form a state that is simultaneously crystalline and liquid at certain temperatures; at higher temperatures, however, this liquid crystal changes into an ordinary liquid. This phenomenon attracted sporadic attention among researchers during the 1940s and 1950s, but liquid crystals were generally considered an area of niche research with little practical value or technical importance well into the mid-1960s. As late as 1967, for example, researchers could confidently write that "whilst liquid crystals are uncommon and of no practical importance, they are of interest for the light they throw on the conflict between order and disorder" (quoted in Gray 1998, 6).

Those who were interested in potential practical applications for LCDs found a less than enthusiastic response to their work within the engineering community. One pioneer in liquid crystal research, George W. Gray, later remembered that "in the years 1960 to 1968, it was . . . difficult to attract support for niche research like liquid crystals, seemingly without importance or technical value" (Gray 1998, 6). In 1960, however, Richard Williams at RCA demonstrated a liquid crystal display (LCD) prototype display device, and by 1964 George Heilmeier demonstrated an improved version of the device at RCA's David Sarnoff Center in New Jersey, using what RCA termed "dynamic scattering mode" (DSM) technology. Two years later, Glenn Brown at Kent State University organized the First International Liquid Crystal Conference. This helped organize the small number of researchers interested in liquid crystals, providing a network for the exchange of information and spurring interest in the topic. A second conference in 1968 excited further interest, and by 1970 practical applications for liquid crystals were being pursued by a substantial number of researchers, with a variety of corporations beginning to take interest. RCA continued to play a leading role in research in LCDs through the end of the 1960s.

DSM technology utilized the ability of liquid crystal molecules to scatter light to display basic letters and numbers. A thin film of liquid crystal material is placed between two layers of solid material, with the top layer being transparent and the bottom layer having a reflective surface. Transparent, conductive electrodes between the sheets made contact with a thin layer of liquid crystal material. Voltage could be applied to one or more of the electrodes to activate a region of the liquid crystal material. In an inactivated state, ambient light is unable to pass through the liquid crystal, which, as a result, appears dark. When an electric field is applied to the crystal, however, the molecules align themselves perpendicular to the field, causing a state of turbulence. This turbulence, in turn, scatters ambient light because of the spatial variation in the index of refraction. Unactivated areas thus appear dark, while activated areas scatter light toward the viewer and against the mirror surface, and appear bright, making it possible to manufacture simple alphanumeric display devices. This process requires very little energy because the LCD modifies ambient light rather than generating its own light, as an LED does. As a result, liquid crystals offered the promise of a viable display technology for use in applications with very low power sources. One important disadvantage of DSM technology, however, is the mirror-like surface it required. Since the image was seen through the reflection of the mirror, the alphanumeric display could appear washed out and could cause eyestrain.

As a result, DSM displays were not easy for people to read and were never widely used. Instead, a variety of other liquid crystal operating modes were developed that sought to avoid such problems. These used "nematic" liquid crystals with twisted structures that untwisted when an electric field was applied. The untwisting blocks light passing through the assembly, making that spot look dark. An array of these spots, each addressed by a pair of electrodes, can form a composite of light and dark spots to form a more complex image, or simpler arrays can be used to form digits or letters. Yet as promising as RCA's liquid crystal displays seemed, they would not become the dominant technology in the market. Instead, new operating modes developed in the following years would propel LCDs to a central role within the field of display devices.

THE LASER

The maser, discussed earlier in this chapter, was an outgrowth of microwave research, specifically the effort to find efficient and powerful tubes to generate or amplify short wavelengths for radar and communication. It was no great intellectual leap to consider moving even further up the electromagnetic spectrum into the submillimeter range, with the goal of producing infrared or perhaps even optical radiation. Accomplishing that was an entirely different matter. Masers could only just reach the centimeter range in the late 1950s, so it was widely assumed that it would be years before submillimeter masers could be considered.

Despite this, in the summer of 1957 maser inventor Charles Townes began to work on developing a maser that would generate waves in the infrared range. While thinking about the frequency of oscillation and its relationship to the "masing" substance, Townes realized that under certain circumstances the maser would allow for the production of waves at much smaller wavelengths than was then possible. Townes later recalled that "I suddenly realized that maser techniques could just as easily be applied to the visible region and in fact visible waves would probably be easier than the far-infrared, because the equations for an oscillating system showed that no more excited atoms or molecules were necessary for a visible oscillator than for a far-infrared one, and techniques in the visible range were already well developed" (quoted in Bromberg 1991, 67). Townes's realization, in other words, offered the possibility of masing action in the optical range, allowing researchers to jump right past the millimeter wave range and go directly to visible light. Townes quickly began intensive research on the topic and, with his colleague Arthur Schawlow, published an important paper in 1958 in *Physical Review* on the concept of the "Laser," which stood for light amplification by the stimulated emission of electrons. Townes and Schawlow would later receive a 1960 patent for the idea of the laser.

Yet Townes had not yet built a working laser. In the years immediately following the publication of their *Physical Review* article, Townes attempted to demonstrate a laser made by stimulating potassium vapor, while Schawlow decided to investigate solid-state materials, initially choosing ruby due to its unique optical properties and easy availability. Both proposed using a bright light source to "pump" the laser and stimulate the emission of short pulses of light.

They were joined by other researchers who investigated other materials, including Ali Javan, who used a combination of helium and neon stimulated by an electromagnetic field rather than a powerful lamp to produce a continuous beam. Another early researcher was John Sanders, who began to work on pure helium. At the same time, a former student of Townes named Gordon Gould also began working toward a laser. Gould would later claim that he had developed his ideas independently of Townes, and the two would become involved in a long-running patent dispute over who had in fact invented the device. In the late 1950s, in other words, there was something of a race to develop the laser.

Most agree that the first laser to be demonstrated was a solid-state ruby laser developed in 1960 by Theodore H. Maiman of Hughes Research Laboratories. The research community as a whole had, by this point, decided that ruby was not a feasible material for laser research. Schawlow, for example, had decided that a ruby laser would require too much power for continuous operation and had abandoned his efforts in that direction. In addition, Irwin Wieder had presented an influential paper at a conference that argued that ruby would only be about 1 percent efficient in a laser device. However, according to George F. Smith, Maiman's former manager at Hughes Research, Maiman simply didn't believe what had been written about the ruby's limited suitability for lasing. He rejected the conventional approach, which focused on the use of gasses, noting that they were difficult to work with, easily contaminated, and often corrosive. Solids seemed more promising to Maiman since they could handle higher power and were more rugged. They could also be operated under a wider range of temperature conditions and could conceivably be used to make devices that were smaller in size. In May 1960, Maiman demonstrated his first experimental ruby laser. Built with a chromium-doped ruby crystal, and pumped by a pulsed, high-power flash-lamp, it operated in the red portion of the light spectrum at a wavelength of 0.6943 micrometers. Surprisingly, Maiman encountered considerable obstacles to publishing his discovery, which was considered "just another maser paper" by the editors of the prestigious *Physical Review Letters*. Instead, he turned to the English journal *Nature*, which gladly took the honor of publishing the first public disclosure of the laser.

With the demonstration of the ruby laser, the research and development of solid-state lasers was rapidly taken up by a wide variety of researchers. By 1960 at least four different types had been demonstrated. Ali Javan, D. R. Herriott, and W. R. Bennett, for example, demonstrated the first gas laser. Javan's device was also the first laser capable of continuous operation. It operated at a wavelength of 1.15 micrometers, and attracted a large amount of interest among other researchers. Many different materials, even uranium, were soon demonstrated to be suitable for lasing. In 1961, researchers at Bell Laboratories built a neodymium laser that combined solid-state design with room temperature operation; a year later the first continuously operating solid-state laser was demonstrated, which used an obscure material known as Sheelite-Neodymium ($CaWO_4$:ND) to obtain the lasing action. However, following the ruby laser, the most important solid-state laser developed during the 1960s was the ND:YAG (neodymium yttrium–aluminum garnet) laser, invented in 1964 at Bell Labs by J. E. Geusic, H. M. Marcos, and L. G. Van Uitert. It emitted up to several hundred watts of continuous power at a wavelength of 1.06 micrometers. This type of laser was the starting point for subsequent research in semiconductor lasers, and also had several important commercial applications. One important successor to the YAG laser, for example, was the "Alphabet YAG" laser of 1966, named because of the large number of doping elements added to the yttrium–aluminum crystal.

The development of gas lasers continued during this time as well. In 1962, for example, Alan D. White and J. Dane Rigden at Bell Labs

Theodore Maiman poses with an early laser, 1960. © Bettmann/CORBIS.

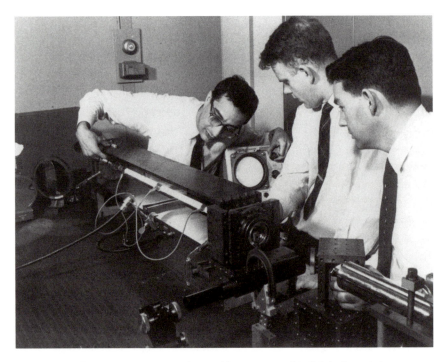

Ali Javan, William R. Bennett Jr., and Donald R. Herriott of Bell Telephone Laboratories adjust a helium neon laser, 1961. Courtesy Lucent Technologies Inc.

demonstrated the first continuously operating, helium–neon laser. By 1964, Kumar Patel, also at Bell Labs, had developed the carbon dioxide (CO_2) laser. The CO_2 laser was the first practical laser capable of both high power and continuous operation. Others, such as Earl Bell and Arnold Bloom at the Spectra-Physics Corporation, experimented with mercury ion lasers. The work of William Bridges in rare-gas ion lasers also generated considerable excitement in the field, and led to krypton-, xenon-, and neon-ion lasers in just a few short years. The most important of these ion lasers was probably the argon-ion laser, accidentally discovered by Bridges in 1964. The argon-ion laser was a major breakthrough because it promised dramatically increased levels of power output. By October 1964, researchers at Raytheon had obtained an output of 4 watts using argon ions, which was an increase of about 1,000 over previous gas lasers. Raytheon scientists achieved 8 watts of output by January of the following year. Further progress in the 1960s included pioneering work in chemical lasers by J. V. Kasper and George C. Pimental at the University of California, and the development of dye lasers by Peter Sorokin and John Lankard. In a chemical laser, the laser pumping occurs through a chemical action rather than photon bombardment. In dye lasers, changing the concentration of dyes in the liquid lasing medium allows the frequency of the light output to be tuned. By the early 1970s, these types of lasers were joined by the excimer (from "excited dimer," a special type of molecule) , X-ray, and free-electron lasers.

SEMICONDUCTOR LASERS

Another important area of research was in the field of semiconductor laser diodes. Laser diodes can be thought of as a type of device combining some of the properties of a gas or solid-state laser and an LED. Some of the pioneering work in this area was done at the Lincoln Laboratories of MIT, where Robert Rediker constructed a diffused gallium arsenide diode that, the researcher learned, could emit infrared light with a high level of energy efficiency. As early as 1960, the Japanese researchers Yasushi Watanabe and Junichi Nishizawa were awarded a patent on a semiconductor maser, and in 1961 the Soviet researcher Basov proposed using p-n junctions in heavily doped "degenerate" semiconductors for laser purposes. Research on gallium arsenide devices would lead to many important innovations, not all of them lasers. In 1962, J. B. Gunn developed what has come to be called the Gunn diode, a gallium arsenide diode capable of oscillation at frequencies of up to 90 GHz, in the microwave range. Gunn diodes remained in large-scale production in the late twentieth century for purposes such as their use in automatic door openers. Another early commercial application for gallium arsenide devices emerged from research at IBM on infrared-emitting diodes. H. Rupprecht, J. M. Woodall, K. Konnerth, and D. G. Pettit published their findings on these devices, made by a new process called Liquid Phase Epitaxy, in 1966. Over 850 million of these diodes were being manufactured yearly by the mid-1990s for a range of purposes, notably television remote controls.

Practical forms of light-emitting semiconductor diodes proved more difficult to construct, but they soon followed. In 1962, Robert Hall at General Electric became inspired by laboratory demonstrations of semiconductor light emission from the Lincoln Labs gallium arsenide diodes, and began to research semiconductor light emission for himself. Hall developed a working semiconductor laser diode using gallium arsenide phosphide by December 1962, which was capable of emitting visible light. The device was a gallium arsenide diode of otherwise ordinary construction, although its ends were polished to form a solid-state resonant cavity to establish the laser action. Hall's design for the semiconductor laser is generally credited as becoming the most influential, and similar devices are used today in DVD players and many other systems.

Others were working on similar projects at the same time. A group at IBM, led by Marshall I. Nathan, for example, also announced the demonstration of a semiconductor laser diode the same year as Hall, while R. J. Keyes and T. M. Quist at Lincoln Labs and Nick Holonyak at General Electric all independently constructed semiconductor lasers at about the same time. Semiconductor laser diodes generated a considerable amount of excitement due to their tiny size—they were about the size of a grain of sand—and high efficiency, and over the next few years substantial efforts were made to improve laser diode technology. Research focused on devices made from GaAs and its alloys, most notably GaAsP, but substantial technological problems soon dampened the industry's enthusiasm. These problems included the need for high current flows in the devices, which led to overheating, and the corresponding difficulties in designing a device that could dissipate heat effectively. The early semiconductor lasers had erratic operating lifetimes and low power-conversion efficiencies. In 1968, researchers at RCA overcame these problems with the development of the single heterojunction laser diode, a multilayered solid-state laser manufactured by the layering of thin, semiconductor films with differing band-gap energies. The RCA laser was made out of GaAs and Aluminum Gallium Arsenides (AlGaAs), still the most widely used materials to make laser diodes through the end of the twentieth century. In 1970, Zoo Hayashi, Morton B. Panish, S. Sumski, and P. W. Foy achieved a major breakthrough by developing a double-heterostructure, higher efficiency laser diode that operated at a much lower current and at a wider range of temperatures than other laser diodes of the time. By 1973, laser diodes were beginning to find use in applications related to thermal encoding, infrared illuminators, and optical communications.

One small index of the intense interest among researchers in lasers during the 1960s is the number of laser-related papers abstracted in *Physical Abstracts* during the period. In 1961 there were twenty; in 1962, 120; and in 1963 there were 270. By the late 1960s, however, there were at least 1,000 papers published every year. Despite the large amount of research and the rapid pace of innovation, however, commercial applications for the laser were slow in coming; the laser, in fact, was sometimes referred to as an invention in search of an application. Even General Electric's exciting solid-state lasers of the early 1960s were unappealing commercially, partly because they were priced at $2,600 each, partly because they could not work continuously at room temperature (if they could have, it would have made them more suitable for communication or other applications). They simply could not compete with existing technologies in commercial applications. GE, in fact, dropped out of the laser business soon afterward. Military applications remained the most important market during the 1960s: Robert W. Ellsworth and R. J. McClung at Hughes, for example, used ruby lasers to invent Q-switching, leading to the development of laser range finding that could be used in tanks and other weapon systems.

However, even from the very beginning, the laser was likened to the "ray guns" of science fiction, and to some this was its future. In 1962, U.S. Army General Curtis LeMay prophesied the use of the laser as an antimissile weapons system. At the time experts dismissed the idea, but the U.S. military still sponsored research on high-powered versions of carbon dioxide lasers through the late 1960s. Power levels as high as 60 kilowatts were achieved, though with the notable exception of Ronald Reagan's later "Star Wars" program of the 1980s, most laser weapon projects were sooner or later abandoned due to technical difficulties.

4

The Peak Years

◆

ELECTRON DEVICES IN TRANSITIONAL TIMES

The period from the late 1960s through the early 1980s was a period of transition in the electron device field. Not always evident from the record of technical accomplishment was the fact that this period was a difficult time in the careers of many electrical engineers in Europe and the United States. Employment issues rose to prominence as the production of many types of semiconductors and tubes shifted to Asia. As a result, the dominance of U.S. manufacturers in the world semiconductor market declined precipitously. Engineering pride was wounded as the American and European consumer electronics markets were increasingly dominated by foreign-produced goods. At the same time, the United States faced a major recession, blamed on military expenditures for the Vietnam War and the energy crisis of the early 1970s. Improved relations between the United States and the Soviet Union, combined with other factors, were good news for world peace but bad news for electrical engineering, because when cuts in government spending combined with the recession, the result was high unemployment in the profession. According to one estimate, as many as 5,000 unemployed electrical engineers lived in the Boston area alone during the hard times of the early 1970s.

With this adversity sometimes came opportunity. The Arab Oil Embargo of the early 1970s contributed to popular support for increased research toward developing alternative sources of energy. Millions of dollars were spent to improve solar cells, which had seen scant attention since the early 1960s. In the United States alone, the federal government committed $1.2 billion to the improvement of "photovoltaic" devices beginning in 1978. However, from the broader perspective this was a minor effort compared to the work that had gone on in the fields of microelectronics and lasers.

A CHANGING WORLD

One small indicator of the expanding importance of non-U.S. manufacturers in the electron device industry was revealed by the shifting demographics of membership in the Institute of Electrical and Electronics Engineers (IEEE). This professional society was founded in the United States and consisted almost entirely of U.S. members until the 1960s. The number of IEEE members had climbed to 200,000 by 1979, but 40,000 of those members lived in countries other than the United States (mainly in Europe). Unfortunately, the growing non-U.S. membership, combined with the economic difficulties of the time, stirred up nationalistic feelings among many American members, resulting in the creation of a new organization called the IEEE-USA, a lobbying organization (the IEEE's nonprofit status did not allow it to lobby directly). The IEEE-USA set out to protect the interests of American engineers and the companies they worked for, a mission that seemed to be in direct contradiction to the IEEE's self-proclaimed role as an international organization.

This expanding non-American membership was the lagging indicator of major transitions in the electronics and electron device industries. Firms in Japan, and later Korea, Taiwan, Singapore, and elsewhere, began manufacturing transistors and integrated circuits, sometimes with devastating results to their competitors in the United States. European tube manufacturers, who had spent much of the 1950s catching up and who had become serious competitors in the 1960s, were just as hard-hit by Asian imports. Some of the new Asian firms were homegrown, but some were created by (or with the help of) established firms in the United States seeking to lower the costs of production. While the general public did not become aware of this transition in the industry until computer memory production became a public issue in the 1980s, American and European firms were gradually squeezed out of the production of commodity products such as ordinary transistors during the 1970s.

Although it is a difficult trend to document, the 1970s can also be seen as a period of transition in device research. Many new devices were invented, including several very important ones such as the free-electron laser and the microprocessor, but the number of breakthroughs did not seem as great as in the two previous decades. There was a considerable amount of technology transfer from firm to firm and nation to nation, but fewer ideas seemed to emerge from the laboratories. Corporate research institutions such as Bell Laboratories were still highly productive, but in retrospect were also in their twilight years. Some institutions, such as RCA's Sarnoff Research Center, would be negatively affected by managerial changes that emphasized applied rather than basic research. All in all, and with very important exceptions, 1970 probably marked the beginning of a period when the most important breakthroughs came in the areas of product development and manufacturing, rather than the invention of entirely novel technologies.

SAW DEVICES

The universe of semiconductor electronics was nonetheless expanding beyond transistors, diodes, and computer chips. Some of these new applications are worth mentioning, but are so distinctive that they are difficult to integrate into the larger story. One such class of electronic device, physically similar to but functionally distinct from the integrated circuit, is the so-called surface acoustic wave (SAW) device. The physics of the propagation of mechanical waves through a solid medium was explored by Lord Rayleigh in an 1885 paper on seismic activity in the earth. Much later, electronics engineers investigated the use of acoustic waves traveling through solids as a way to manipulate radar signals. Radar researchers had discovered that their radars worked better if they, in effect, lengthened the outgoing radar pulse. However, that required that the receiver delay the reflected, incoming pulse. To create the delay, they used a piezoelectric transducer, a mineral crystal material that vibrates when subjected to an electric field, or that generates an electric field when subjected to vibration. With one transducer transmitting vibrations through the medium, the other received them and converted them back to electric signals after a slight delay. By carefully designing the medium in a particular shape, different "slices" of the signal were delayed by differing amounts, and when recombined this resulted in a complete but "compressed" signal. Some of the fundamental work was undertaken in the years before 1962 by John H. Rowen, Erhard K. Sittig, and Richard M. White, among others.

Until the 1970s, nearly all SAW devices were used for such pulse compression applications. Later studies of wave propagation led to the identification of many different materials suitable for use in electronic SAW devices, such as lithium niobate. At the same time, engineers began to develop new pulse compression filters, bandpass filters, oscillators, and other devices that used SAW technology. A major breakthrough in terms of the commercialization of SAW devices resulted from the work of Robert Adler at the Zenith Corporation, whose team developed SAW devices for use as intermediate-frequency (IF) filters in television receivers. Earlier types of IF filters for television were based on a vacuum tube and related circuitry, but the new SAW IF filters were very small, were inexpensive to produce, and did not require adjustment at the factory. They could be manufactured in bulk using photolithography and other integrated circuit techniques, and housed in inexpensive plastic packages. As a result, they soon became a standard component in television sets. SAW filters also began to be used in other communications gear during the 1970s, such as filters for television broadcasting equipment and digital radio communication. The 1970s also witnessed considerable effort to integrate SAW devices and integrated circuits, a natural development given the considerable amount of ancillary electronics needed to use SAW components. However, by the late 1990s this technology was still not available in practical form.

Looking forward to the end of the twentieth century, hundreds of millions of SAW devices would be produced every year to be used in a wide range of equipment, from radar and radio gear to television receivers, VCRs, pagers, and cellular telephones. Though little-known compared to famous cousins like the microprocessor, the SAW device had quietly become one of the most widely used types of electron device.

SEMICONDUCTOR MEMORIES

The major commercial breakthroughs of the 1970s were semiconductor memories and the invention of the microprocessor. Both innovations had been nearly ready by the late 1960s and began to be produced in large numbers in the early 1970s. Both signaled the beginning of a transition in the device field from high-flying research toward down-to-earth commercial applications. The programmable computer became the basic model for the industry as a whole, and it is perhaps not too large of an exaggeration to mark this period as the beginning of the "information age."

The electronics industry had developed various forms of magnetic memory storage by the mid-1950s, including magnetic tape and magnetic drums. There was considerable interest in computer memory modules made from grids of small, magnetizable rings called "cores." A binary "one" or "zero" could be stored in a core as a magnetic field, aligned in one direction or another. The cores were magnetized or demagnetized by wires, running through the core's center hole, which intersected at right angles. Then, once a bit was "written" to the core, a sensing wire was used to detect the state of a core's magnetization. Large numbers of cores were woven into a textile-like matrix of wires to create a memory unit. Given the rapid progression in computers from vacuum tubes to integrated circuits in the 1950s and 1960s, it seems surprising that such a system was considered the most cost-effective type of memory as late as the 1970s, but that is in fact the case. There were many attempts in the 1960s to use transistors (either discrete devices or those fabricated as integrated circuits) for computer memory. Benjamin Agusta, R. D. Moore, and G. K. Tu announced such a chip, using bipolar transistors, in 1969.

However, with the emergence of MOS integrated circuit technology in the late 1960s and the higher circuit density it allowed, practical semiconductor memory became feasible for the first time. The first semiconductor memory chips began to appear soon after MOS technology made its entrance, and were commercially available by the end of the decade. It took several more years for them to displace the entrenched technology of cores. One of the first semiconductor memory devices to make use of this new technology was the read-only memory (ROM) semiconductor memory chip. This type of memory contains fixed instructions and other data needed to run a computer, and because it never needed to be updated, it was designed into the chip's circuitry, making it "read only." Fairchild designed one such chip as early as 1967, consisting of a MOS integrated circuit of 64-bit capacity. The next year, Philco-Ford offered a 1,024-bit ROM chip. Random-access memory (RAM), the kind used to store data or programs temporarily, was developed in MOS form toward the end of the decade as well, and in 1970 Fairchild offered a 256-bit RAM chip. That same year, Robert E. Kerwin, Donald L. Klein, and John C. Sarace at Bell Labs introduced a new way of making MOS transistors with self-aligning gates. This further reduced the minimum size of the transistors and became one of the basic processes used in MOS chip construction for many years.

A basic distinction between two types of RAM appeared almost immediately: dynamic RAM (DRAM), which requires only one transistor for each bit stored but which can retain the data for only a short amount of time before needing to be renewed, and static RAM (SRAM), which requires six transistors per bit but can store data for as long as desired. Dynamic RAM was seen as the key to future developments, both because it allowed larger amounts of data to be stored on a single chip and because it was less complicated, and therefore less expensive, to manufacture. The excitement over RAM chips was fanned by the announcement of new types of MOS devices, such as high–performance vertical–groove MOS (VMOS), invented by T. J. Rodgers at Stanford University in 1972.

Electronic memories generated the kind of entrepreneurial free–for–all that had characterized transistor manufacturing almost two decades earlier. In 1968 Gordon Moore, Robert Noyce, and Andrew Grove left Fairchild to found a new company named Intel. Moore, Noyce, and Grove started Intel with the purpose of producing semiconductor memory, and in 1971 they introduced the first commercially successful 1024-bit (1-Kbit) dynamic RAM chip (relying heavily on Bell Labs' self–aligned gate technique, developed further at Fairchild). This was the device that truly established semiconductor memory as a serious competitor with magnetic core memory technology. Another significant innovation at Intel during the same period was the development of programmable read–only memory (PROM), which can hold information supplied by the user for as long as the user desires. PROM chips were valuable to manufacturers of complete systems, such as calculators or computers, because the programming in a PROM chip could be set after the chip was manufactured, unlike a standard ROM chip.

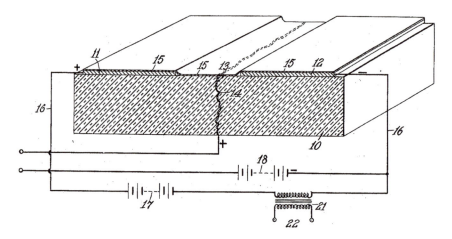

Julius Lilienfeld's 1930 patent for a device much like a metal oxide semiconductor (MOS) transistor. His device used two glass plates or blocks, each with a gold or silver terminal (11 and 12), between which was held a thin strip of aluminum foil (13). A thin film of semiconducting metal was sprayed or coated on the surface, touching the terminals and the edge of the foil. The electrostatic field generated by a current applied to the terminals affected the conductivity of the semiconductor layer, so that the device functioned analogously to a vacuum tube. U.S. Patent 2994018.

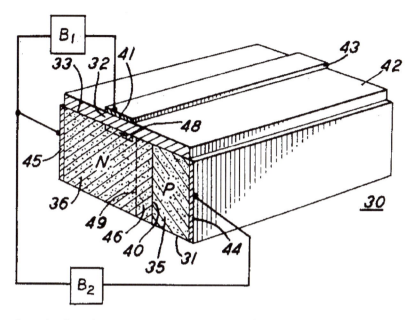

Atalla and Kahng's basic MOS transistor consisted of a semiconductor junction and an electrode (43) separated by a silicon oxide insulating layer (42). U.S. Patent 3206670.

BIGGER MEMORIES

As circuit densities rapidly increased due to advances in MOS technology, memory capacities likewise increased. By 1972, many companies were producing memory chips of 1K capacity. In 1973, 4K RAM chips were introduced. Manufacturers offered the first 16K RAM in 1976–1977, and then shifted to 64K DRAM and 16K SRAM in 1980–1981. Coupled with this increase in performance came drops in prices. To cite one example, in 1980 a 16K dynamic RAM chip cost about $2.50. A year later, the same chip sold for about 90 cents. This rapid increase in power and equally rapid decrease in price meant that instead of designing semiconductor memory chips for specific applications, applications instead were designed with the idea of semiconductor memory in mind. Further, systems designers were quick to embrace the idea of using more and more memory. Cheap memory led to computers with ever-larger memory capacities, and this led to new, memory-intensive software applications. Whereas before software had to be written as if memory were at a premium, with every passing year that restriction became less and less important. This led to what one observer has referred to as an "insatiable market" for semiconductor memory devices during the late 1970s and 1980s.

Paralleling developments in memories were advances in logic. Chips based on complementary MOS (CMOS) designs, first announced in the early 1960s, would finally come to fruition, leading to faster computers. However, not everyone was convinced that MOS technology was the best route. As engineers began to discuss large-scale integration as the next logical step in chip design, they investigated new technologies such as integrated injection logic (IIL), announced in 1972, to provide chips with 1,000 or more logic gates. This modification of the standard procedures used to make bipolar transistors produced high-speed circuits with features as small as 5 micrometers. While logic chips remained in production, their evolution was completely overshadowed by the appearance of a remarkable new device called the microprocessor.

THE MICROPROCESSOR

The second crucial development for the semiconductor industry during this period was the invention of the microprocessor in 1971. In 1969, Intel had been approached by the Japanese calculator firm Busicom and asked to manufacture a set of integrated circuits for use in a new line of desktop calculators. Each of the calculators had somewhat different features, and presumably would require a different set of operating instructions to be designed into its chips. One of the employees at Intel, Ted Hoff, realized that instead of designing a different set of chips for each of the different calculators Busicom intended to produce, he could design a single, programmable chip set that could be used in them all. Before this, calculator chips were always designed to manipulate data in certain specific ways in order to produce the desired output. The detailed functions of the finished machine had to be specified in advance so that the chip could be designed accordingly. The same held true for logic circuits in general: each was designed to manipulate data in certain specific ways depending on the application in which it was to be used. As a result, as applications became more and more complicated and required larger and larger numbers of logic circuits, the time spent in designing chips mushroomed. The industry faced a serious bottleneck problem as a result of its own success.

The invention of the microprocessor helped solve this problem. With the microprocessor, there is an extremely large number of possible routes for the data to take built into the device, but a set of instructions fed to the chip temporarily determines the way the data will be manipulated. In other

A microphotograph of the Intel 4004 microprocessor, suggesting the microscopic dimensions of the devices and interconnections. Courtesy of Intel Corp.

words, where the earlier type of calculator chip had a limited number of data routes hard-wired into it, from which the user could choose, the microprocessor has an extremely large number of possible routes within which a subset is determined by the device's programming. Such a programmable processor would be more complex than a special purpose chip, and it would require that extra software be written to control it, but that software could be adapted as necessary to different systems, giving the chip a high degree of flexibility. It was, in other words, an analogy to the computer system itself, which could be programmed to do a wide range to tasks. The power of Hoff's insight was that it allowed a single chip design to be used in an extremely wide variety of devices from calculators to small computers to industrial process controllers, depending on the needs of the customer. As a result, separate logic circuits did not need to be designed at all—a single, universal logic circuit could be used instead.

Gary Boone on the Invention of the Microcontroller

Gary Boone was an engineer at Texas Instruments when he designed the first "computer on a chip" microcontroller.

The Texas rule that applied was "one riot, one ranger." That is, one chip, one engineer. So TI, with maybe twenty engineers, can deploy three or maybe four of these project teams at any one time. It takes maybe six months to do one, and so that's the capacity of this business. It's the number of engineers divided by the number of chips every six months. Plus, speaking as one of the engineers who got put on those teams, they all looked pretty much the same. The individual requirements differed in detail, but in principle and in overall function, they were almost identical. So, what goes through your mind is, "I'm tired of doing this. I'm working long hours. My family is not happy. I've got to find a better way to do this." You end up thinking about a sort of a matrix of customer requirements one way and functional blocks or chunks of circuitry the other way, and you identify the commonality and you mentally consider and sort of simulate, "Okay, now if I had this many bytes of data storage, and I had this many bytes of program storage, and I had this many bits of keyboard scan interface, then that would cover all of the specifications I know about, maybe. . . ." So that's the genesis of the TMS 100 microcontroller chip: it came out of boredom, high demand, and a vision of commonalities that were being inefficiently served by deploying these huge teams with so many chips.

Source: Gary Boone, an oral history conducted June 22, 1996, by David Morton, IEEE History Center, Rutgers University, New Brunswick, New Jersey.

INTEL'S 4004

Intel announced the world's first microprocessor, the 4004, on November 15, 1971. Designed by engineer Federico Faggin, it was capable of processing information in 4-bit length (4 binary digits), performed about 60,000 binary operations per second, and used approximately 2,300 transistors, more than twice the number common for calculator chips at the time. The 4004 was not offered as an independent product, but was sold as part of a larger set of chips that formed the basis of an entire microcomputer system, the MCS-4. This system consisted of a 256-byte (a byte being a string of bits, usually 8, representing a single number or character), type 4001 ROM chip; a 32-bit 4002 RAM chip; a 10-bit, type 3003 shift register chip; and the 4004 microprocessor. Clock speed on the CPU, a general indication of how fast the chip performed its duties, was 108 kHz, and the initial price was $200. The following year, Intel moved to integrate even more of the functions of this system onto a single chip, rather than offering a chip set. The result was the first commercial 8-bit microprocessor, the 8008, first offered in April 1972. It used 3,500 transistors, could perform 60,000 operations per second, operated at a 200 kHz clock speed, and could access 16 kilobytes of memory. By 1973, Intel offered the 8080, a modified version of the 8008.

The development of the 8-bit 8008 and 8080 chips was important in part because it allowed for more complex alphanumeric calculations, something impractical with a 4-bit system. It served as a model for the rest of the industry, and soon spawned a host of competitors from manufacturers such as National Semiconductor, AMI, Motorola, and Fairchild. By the mid-1970s there were approximately forty microprocessors on the market, from virtually all the major semiconductor companies. Motorola, for example, introduced its 6800 chip in 1974, an 8-bit microprocessor designed for use in microcomputers and industrial controllers and the automotive engine management computers then coming on the market. Around the same time, a variety of companies introduced the first generation of 16-bit processors. Intel, for example, introduced its 8086 chip in 1978, a 16-bit, general purpose microprocessor that contained over 29,000 transistors.

THE PERSONAL COMPUTER

The most memorable result of this rapid period of innovation was the emergence of the personal computer. Personal computers of one sort or another had been prophesied since at least the early 1950s, and in fact *Radio Electronics* magazine published a series of articles on building a home computer in 1950 and 1951. The Heathkit Corporation, a maker of kits that hobbyists assembled into hi-fi and radio gear, offered a simple analog computer kit in 1959. Digital Equipment Corporation offered a "desktop" version of its popular PDP-8 computer in 1968, and although it was far too expensive for the average consumer and it came with little or no software, it inspired a number of people around the world to form clubs devoted to using computers. Once the PDP-8 and others had led the way by using inexpensive integrated circuits, a number of other small, home-built, or desktop computers appeared in the early 1970s. In fact, it is probably impossible to identify any of these as being the absolute first personal computer. For many people, however, the "Mark 8" computer stands out as a milestone. This small computer, with no keyboard or monitor and programmed entirely by flipping a set of front-panel switches in a certain sequence, was announced in *Popular Electronics* magazine in 1974. It set off a flurry of hobbyist interest that grew even more the next year, when *Popular Electronics* ran a "how-to" construction article on a second personal computer called the Altair. By 1976, when the first Apple II computer was offered to the public, the personal computer hobby had gained considerable momentum. In quick succession in the late 1970s and early 1980s, IBM, Commodore Business Machines, Radio Shack, the Atari Corporation, and others introduced small, inexpensive computers for use in home and office. Advances in sheer power predictably continued: the first 32-bit microprocessor, for example, was introduced in 1981. It used more than 200,000 transistors and was built on three chips.

FROM MICROPROCESSORS TO CONTROLLERS

The dizzying increases in the complexity of cutting-edge microprocessors was only part of the story. Less advanced, 4-bit microprocessors continued to outsell more powerful devices during this entire period, as designers integrated them into a wide variety of applications that did not demand high levels of computing performance. There was a huge potential market beyond the few thousand systems sold every year to the military and to large businesses and universities. One of those markets was industry, where automated production had taken hold as early as the 1950s. A class of systems called industrial controllers, simple programmable "black boxes" to control various kinds of industrial processes, simply begged to be redesigned as microprocessor-based systems. The first microcontroller, a type of microprocessor developed for this market, was designed by Gary Boone and Michael Cochran at Texas Instruments in 1971. The difference between the microprocessor and the first microcontroller was that the microprocessor was for the most part limited to processing information supplied to it from external sources, such as memory chips, while the microcontroller included many of those peripheral functions of a computer on the same chip as the processor, including input and output capabilities. In some sense, then, the microcontroller was the real "computer on a chip."

In 1972 Texas Instruments released its TMS 1000 4-bit "microcomputer" (microcontroller) chip, and by 1979 over 26 million of these devices were being sold annually. This widespread popularity, which overshadowed sales of microprocessors, was due to the fact that a single-chip microcontroller could be integrated into various kinds of systems more easily than microprocessors. As a result, an extremely wide variety of technologies soon began to appear that made use of these single-chip computers, including automobile engine management systems, industrial devices, scientific calculators, home appliances, and even toys. A perfect distinction between the microprocessor and microcontroller is not always possible, of course, since many microprocessors also contain component devices that serve "peripheral" functions, and because a microprocessor can be used in many of the same applications as a microcontroller, but the definition remains useful and is, in fact, still reflected in patent law. Four-bit microcomputers, whether based on "microprocessors" or "microcontrollers," have become crucial components in a vast number of commercial, military, and scientific applications. Gary Boone estimates that today approximately 2 billion are consumed per year, a truly incredible number.

Robert Rediker on the Names of Things

Robert Rediker is one of the coinventors of the GaAs laser.

The Gunn effect was a way to transfer electrons from one conduction band valley to another. It operated as an amplifier oscillator. Mr. Gunn at IBM invented it. Everybody jokes that the secret of success is, "If you say at the beginning that it cannot be explained any other way, it gets named after you." In the way the Esaki diode could be called a tunnel diode, the Gunn effect could be called a transferred electron amplifier, but it's always been called the Gunn effect. Some important effects in physics are named after people because no one really understood what they were. That's humor. I've had very delightful people tell me that the major factor in having effects named after them was "because I had a friend."

Source: Robert Rediker, an oral history conducted July 27, 2000, by David Morton, IEEE History Center, Rutgers University, New Brunswick, New Jersey.

THE TRANSFORMATION OF IMAGING AND DISPLAY DEVICES

While the integrated circuit field was dominated by the commercial success of the microprocessor, which tended to reduce the range of specialized computer chips, in the field of imaging and display the number of new devices increased. As early as 1972, one observer wrote that

> selecting a small numeric or alphanumeric display was, until a few years ago, a relatively easy task. There simply was very little choice. Now the situation has changed. The proliferation of display technologies has provided an embarrassment of riches. (*IEEE Spectrum*, 1972)

This increase in commercially available display devices included a wide range of gas-discharge devices, LEDs, and LCDs, in addition to new technologies such as light-emitting film and electrophoretic displays. Despite the assumption that these types of newer technologies would replace CRTs in short order, the market position of the cathode ray tube was not seriously threatened by the development of either LEDs or LCDs during the 1970s. In fact, improvements in CRT resolution, contrast, and ruggedness continued to be made throughout the decade, and CRTs continued to be an extremely popular form of information display technology, especially for television, oscilloscopes, and radar. In fact, the CRT was about to receive a new lease on life with the advent of the personal computer.

Newer display applications, however, did tend to use the thin-panel displays made possible by advances in recent technologies and, eventually, the increased utilization of digital information-processing techniques. As a result, LEDs, LCDs, and gas-discharge panels became standard technologies for use in a rapidly growing list of applications during the 1970s requiring small and simple displays, such as watches, calculators, and handheld test equipment.

New gas-discharge displays were introduced by several companies during the 1970s, including Burroughs, Owens-Illinois, and Nippon Electric Company. Like the earlier technology of fluorescent lighting, gas-discharge devices depended on the fact that certain gasses glow when they are stimulated by an electric field. The Burroughs Panaplex was typical of these new types of displays, being a "sandwich" containing thick-film electrodes that acted as cathodes, a spacer frame, and a front glass cover containing transparent anodes. By 1975, gas-discharge displays were widely used, mostly in applications such as the numeric readout in desktop calculators. They offered low production cost and high reliability, and were available in sizes larger than those possible with LED or LCD displays. By 1978, large arrays of gas-discharge devices were being used in place of CRTs in several types of devices, including computer monitors, and were, in the words of one observer, "no longer a laboratory curiosity, but a viable alternative to the cathode-ray tube in several applications" (Torrero 1978, 78). However, one serious problem with gas-discharge displays was that they needed a fairly high voltage of about 170 volts to operate, and therefore required relatively large and expensive power supplies similar

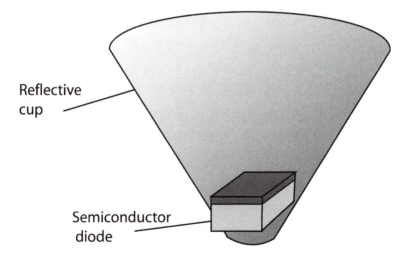

A typical light-emitting diode (LED), showing the semiconductor diode structure mounted in a reflective cup (shown in cutaway view). Light is emitted at the junction and redirected by the reflector.

to that used in vacuum tube equipment. This also made them impractical for portable or battery-operated equipment unless fitted with expensive voltage converters. Furthermore, the high voltage necessary had a tendency to interfere with the workings of nearby integrated circuits, presenting problems for those designing computers and other sensitive equipment. As LED and LCD displays became cheaper, larger, and more efficient, the high voltage required by gas-discharge displays thus seemed increasingly prohibitive.

IMPROVED LEDS

In 1972, Sony Corporation introduced a new method for producing GaP LEDs called synthesis solute diffusion (SSD). This lead to higher efficiency rates and lower production costs. In SSD, crystals of GaP are grown under a phosphorus vapor pressure of about one atmosphere and at temperatures of around 1200 degrees Fahrenheit. Crystals as large as 47 millimeters in diameter and as heavy as 170 grams could be produced using this method, with high-efficiency junctions formed through a new liquid epitaxy process. Green and red LEDs could be produced cheaply and reliably using SSD. The same year, M. George Craford of Monsanto Corporation invented the yellow LED, and while it would take some time to commercialize it, it was an important addition to the LED family. Vapor-phase nitrogen doping of GaAsP (gallium arsenic phosphide) was also developed in the mid-1970s, leading to the production of LEDs in various colors and with high output-intensity levels. Liquid-phase growing of GaP LEDs, developed at Bell, produced even brighter and more efficient green LEDs, and by 1977 efficient LEDs in red, orange, yellow, and green were commercially available for use in a wide variety of applications, including instruments, calculators, and watches.

LCDS

LCD devices also became increasingly important during the 1970s after the invention of the twisted nematic (TN) LCD by two researchers at the Swiss firm Roche, Martin Schadt and Wolfgang Helfrich, and almost simultaneously by James Fergason at Kent State University. Several former RCA employees formed a new firm, Optel, which moved quickly to commercialize a DSM LCD based on RCA's version of the technology. But TN LCDs quickly became the standard for use in applications such as watches, calculators, and other battery-powered devices, replacing the less energy-efficient LED. Sharp Corporation in Japan became the first major electronics manufacturer to offer a calculator with an LCD display in 1970. Long operating life and reliability were other benefits to liquid crystal technology, and between 1975 and 1977 the average LCD's operating life had increased from about 15,000–20,000 hours to 50,000 hours. Despite these improvements LCDs were still suitable only for small, simple, and relatively static displays throughout the 1970s, because in larger arrays the response time (the time needed for the image to change) was simply too long for applications such as television. Further, it was difficult to construct a display comparable to a television screen, with thousands of individually addressable pixels, because each pixel had to be contacted by conductors that all led to the outer edges of the display. There was progress in working around these limitations, but engineers had far to go. In 1978, for example, a 2-square-inch display built by Hughes Electronics (by suspending the liquid crystal directly over the silicon circuit used to activate the pixels, thereby speeding it up) was considered a notable achievement, while CRTs measuring more than 30 inches diagonally or more had been in production for some time.

THIN-FILM TRANSISTOR DISPLAYS

The thin-film transistor (TFT) is an alternative way to construct various types of transistors and diodes using thin films of material, deposited on a substrate, rather than being etched as with other devices. RCA engineer Paul Weimer and his team demonstrated the process as early as 1960, when they constructed FET transistors built up from layers of metal and a semiconductor material such as cadmium selenide. The performance was not quite as good as that of a silicon transistor, but close. By 1963, Weimer was working on a solid-state camera device based on the TFT, which was a natural avenue for television leader RCA to pursue. Working at Westinghouse, T. Peter Brody continued Weimer's work, discovering a wide range of materials that could be used to make TFTs—even strips of paper. Brody believed that the TFT would lead to important applications in the display field. He concentrated on improving the Westinghouse line of electroluminescent displays, using TFT arrays to construct a thin, flexible, and nearly invisible matrix of transistors to control each pixel in a large display. When LCD displays were introduced, he transferred this same "active matrix" idea to that. Unfortunately, the direction of display technology was shifting toward silicon, and most TFT research died out in the 1980s. It enjoyed a big comeback in the 1990s, however, after flat-panel computer displays had become established and as manufacturers were looking for ways to improve monitors and televisions. Early in the history of the LCD display, engineers recognized that a simple array of liquid crystal elements could not provide a "high-motion" moving image with adequate contrast to satisfy the requirements of television. What was needed was an active device such as a transistor at each pixel to act as a high-speed control. Following Westinghouse's exit from TFT research in the 1980s, several Japanese firms led the way toward improved LCD-TFT displays, beginning with Seiko's 1-inch screen LCD television of 1990. Since that time, the active matrix LCD display has become a standard feature of personal computers and small televisions.

WATCHES AND CALCULATORS

LED and some LCD devices were first used in scientific equipment, but consumer applications were where they made the most dramatic impact during the 1970s. One important application already mentioned was the electronic calculator, which employed a number of technologies over the years including gas-discharge, LED, and LCD displays. Introduced around 1970 by Hamilton Watch, the "Pulsar," the first electronic digital watch, was housed in a gold case, had a gold band, and was priced at about $2,000. It used an integrated circuit to keep time and a red LED array to display the hour and minute. Prices for electronic watches began to drop dramatically immediately afterward, and Time Computer soon offered a stainless steel model for the relatively inexpensive price of $275. But prices did not stop falling there. By 1975 there were approximately forty-five different manufacturers of electronic watches, and in 1976 Texas Instruments introduced a model that sold for only $19.95. In 1976 Texas instruments also introduced the first electronic watch that utilized a LCD, and although this watch was more expensive (priced at between $275 and $325), prices again quickly dropped. By the 1980s, basic electronic watches had fallen in price below the level of the most inexpensive mechanical watches, and were regularly given away free in breakfast cereal boxes, sold in gumball machines, or offered for a few dollars in the most common retail establishments. No more public demonstration of the effects of mass production could be made, unless perhaps one looked to the electronic calculator.

Like watches, calculators contained much more than just display devices. They had, for example, been the first household products on the market incorporating microprocessors. But the displays used on calculators were a significant part of their cost, and one that was affected by changes in technology. LEDs and LCDs had been preceded by small gas-discharge displays, although these were more expensive and consumed too much of the power of battery-operated calculators. In 1972, the Monsanto, Sears, and Hewlett-Packard companies all introduced small "pocket" calculators that used LED displays and were priced at slightly less than $100. At this price level, the calculator was inexpensive enough for home-based businesses, students, and accountants to afford. In the same year, Heath Company introduced a desktop calculator that used a Sperry gas-discharge display, and Texas Instruments introduced calculators that used both LED and gas-discharge displays. The number of calculators on the market increased rapidly and prices dropped throughout the decade. The more powerful calculators began to make inroads on the slide rule, that icon of the engineering world, although more advanced "scientific" calculators remained unavailable or too expensive for some years to come.

LEDs and LCDs were also used in other commercial applications as well. General Electric, for example, developed an all-electronic set of instruments to measure jet engine parameters that replaced earlier electromechanical instruments. Single LEDs rapidly became the technology of choice in a huge range of systems where simple indicators (such as "power on") were needed. Where extremely simple graphical displays could suffice, a row of LEDs became a common sight. But despite such applications, electronic watches and calculators were by far the most commonly used devices to use the more complex alphanumeric LED arrays, and nearly all LCD displays were either alphanumeric or capable of low-resolution graphics.

BEYOND THE LED AND LCD

In addition to gas-discharge displays, LEDs, and LCDs, the 1970s also saw the introduction of a variety of other electronic display technologies. Electrophoretic image displays (EPIDs), for example, were developed in the early 1970s by the Japanese firm Masushita Electric Company, Ltd. EPIDs utilize the electrophoresis (separation by the application of electricity) of bits of white pigment particles suspended in a dark liquid. This mixture is sandwiched between a pair of electrode films, one of which is transparent. An applied current causes the transparent electrode to become either positively or negatively charged, which in turn either draws the particles toward it or pushes them away from it. This causes the display to appear black or white. An array of electrodes can be used to create a more complex image.

Also in the early 1970s, Sigmatron Inc. developed light-emitting film

(LEF) technology. LEF technology makes use of the fact that certain poly-crystalline phosphors give off light when an electric field is applied to them. Using integrated circuit construction techniques, thin layers of zinc sulfide doped with manganese (ZnS:Mn) are built directly onto a light-absorbing substrate. An anode generates electrical pulses that activate the phosphor, which then appears bright against the dark, light-absorbing background. Once the pulse is completed, the phosphor continues to glow but slowly dims, as in a CRT. They were relatively inexpensive to produce, and by 1972 were available for less than $0.75 per digit. Both LEF and EPID technology became increasingly popular throughout the 1970s and were used in a variety of commercial applications. Despite this, neither one came close to matching the commercial importance of LEDs or LCDs, and neither attracted much public attention.

Another notable development was early work toward flat-panel displays based on thin-film technology, LCDs, and other display techniques. RCA researchers had earlier attempted but abandoned the effort, and it was taken up at Westinghouse in the late 1960s. There, researchers won a military contract to develop a 6-inch electroluminescent display using thin-film technology, and another military award to develop a new LCD display. By the early 1970s, Westinghouse engineers, including Peter Brody, demonstrated an electroluminescent display panel using thin-film technology. However, a few years later Westinghouse abandoned this effort, leaving further development of active matrix displays to other firms. Thin-film electroluminescent display matrices were improved by the short-lived California firm Simatron, and introduced commercially by Sharp in 1978. Other firms jumped back into this area, including Tektronix, leading to the commercial development of improved active matrix displays in the 1980s. These and similar displays were aimed squarely at the computer monitor and television receiver markets, but it would be many more years before the CRT relinquished its hold there.

THE CHARGE-COUPLED DEVICE: A NEW IMAGE SENSOR

Image- and light-sensing technologies also continued to be developed during the 1970s, including a large amount of work done in the area of photodiodes. In 1970, for example, Charles A. Burrus and W. M. Sharpless developed the first planar germanium photodiodes, and in 1979 Burrus, A. G. Dentai, and T. P. Lee developed a back-illuminated p-i-n junction photodiode (a diode with a layer of "intrinsic" or undoped silicon between its p-type and n-type layers) from InGaAsP (indium gallium arsenide phosphide) in which the illumination enters the InP substrate of the n junction.

But the most important event in terms of image-sensing technology during the early 1970s was the development of the charge-coupled device (CCD). This is a type of semiconductor device that was originally designed to be a type of computer memory chip called a "shift register." It was first conceived by Willard S. Boyle and George E. Smith of Bell Laboratories in an hour-long discussion in 1969.

The construction of a CCD memory chip involved MOS techniques, but the thousands of MOS-like devices on the wafer did not perform the same function as transistors on an ordinary memory chip. Instead, the device was constructed as an array of silicon capacitors. The capacitors were charged by surface electrodes and detected by another set of electrodes. Each tiny device could also transfer its charge to its neighbor at high speed, making it possible to move large amounts of data into and out of the chip quickly. But it was not the CCD's destiny to be a computer memory chip at all. Continued improvements in MOS memory technology negated the CCD's minor advantages over then-current designs.

The inventors knew that the "injector" for the charge could be in the form of electrodes built into the chip, but if the CCD was put in a transparent plastic case, the injection could be supplied by photons striking the surface. In this case, the tiny MOS-like devices responded to light in a way similar to a solar cell. Almost immediately, the inventors realized that they had a new type of image sensor. In fact, CCDs can be seen as the first true challenger to vacuum tubes in television and video camera applications, although it would be several more years before their impact in this area was obvious.

By the mid-1970s, CCDs were beginning to be used in a variety of commercial applications, including television camera systems, analog filters, and pattern recognition devices. In 1974, the first commercial imaging CCD was produced by Fairchild Electronics, with an array of 100×100 pixels, and in 1975 the first CCD TV cameras were used in commercial broadcasts. The use of CCD technology had many benefits over tube technology for commercial image-sensing applications, including increased durability and sensitivity to light, smaller size, and the possibility of much sharper images. In 1976, for example, Fairchild Camera and Instrument Corporation announced a 244-line–resolution camera sensitive to light at levels as low as a 0.000125-foot candle. Designed for military and industrial applications, it sold for approximately $4,500. Scientific projects were also important in the continued development of CCD technology. In 1979, for example, RCA developed a sensitive 320×512 pixel CCD, cooled by liquid nitrogen to improve its performance, for use in a telescope at Kitt National Observatory. In later years, the development and funding of space technology have continued to be important stimuli for CCD development. Space missions launched in the mid–1970s, such as the Voyager missions, used conventional television tubes. Since 1974, however, NASA has invested heavily in CCD technology, beginning a program to increase the size of CCD arrays and to lower their readout noise levels. By 1978 CCD arrays of 500×500 pixels had been produced with substantially lowered noise levels. NASA continued to invest in CCD technology in the following years, achieving huge 800×800 and 1024×1024 arrays in the next decade that were used on the Galileo mission (1989) and the Hubble Space Telescope (1990).

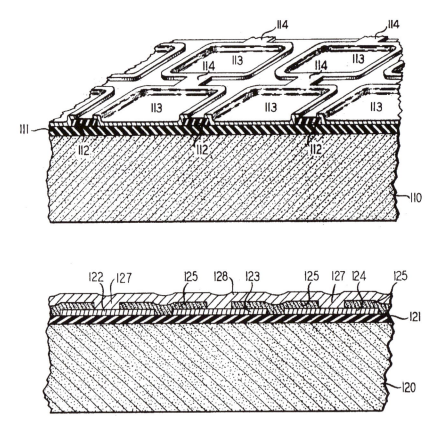

Two views of the original charge-coupled device, showing the construction of its capacitive regions. U.S. Patent 3858232.

The CCD's coinventors, Willard Boyle and George E. Smith, pose with a CCD-based camera. Courtesy Lucent Technologies Inc.

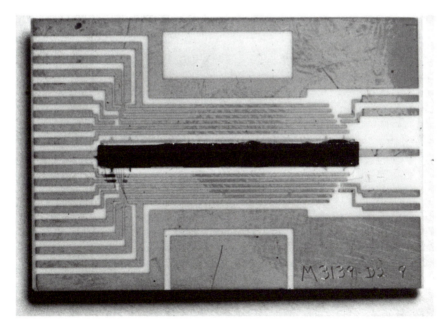

A CCD imaging device made by Bell Telephone Laboratories, 1975. Courtesy Lucent Technologies Inc.

THE LASER IN THE 1970S

The variety of distinct types of lasers increased in the 1970s as it had in the 1960s, though perhaps not quite so dramatically. One important advance, the free-electron laser, was found by Stanford University's John Madey. He began his theoretical work on the topic in 1971 by analyzing the radiation emitted by an electron beam moving through a magnetic field. He discovered that the electrons of the beam responded to the magnetic fields' quantum status as a series of long-wavelength photons by scattering the field into a burst of real, short-wavelength photons. By passing the electron beam through an array of magnets with alternating polarities (a system known as an "undulator" or "wiggler"), the beam could be induced to release a stream of radiation. In 1972 Madey and his coworkers began to build a laser based on this discovery, and in 1976–1977, after numerous setbacks, they finally succeeded in demonstrating radiation amplification at a wavelength of 10.6 micrometers—still not visible light but getting closer. The following year they demonstrated true laser action using this technique, achieving a power level of 7 kW at the much shorter wavelength of 3.4 micrometers. The value of this new laser was that, unlike ordinary lasers, it was "tunable." It could, in other words, emit light at a range of frequencies determined by the design of the equipment. Its tenability and high power led to its being used widely in scientific research and medicine (where it could be used to burn away tumors and make incisions).

Excimer lasers were also developed during the 1970s. An excimer is technically a molecule composed of two identical atoms, which is stable while it is held in its excited state. If the atoms are detached from one another, energy is released. A similar molecule made up of two different atoms, such as xenon chloride, is technically called an "exciplex," but in terms of laser technology the distinction between the two quickly became blurred, and lasers using both types of molecules to produce radiation are referred to by the term "excimer." A group of Russian scientists led by Nikolia Basov demonstrated the first excimer laser in 1970. Interest spread to the United States, where in 1974 Don Setser at Kansas State University produced laser emission in xenon fluoride. In 1975, Stuart Searles at the Naval Research Laboratory demonstrated the first rare-gas excimer laser, using xenon bromide, and a few weeks later James Ewing and Charles Brau succeed in demonstrating lasing action using krypton fluoride at a wavelength of 354 nanometers. Ewing and Brau passed an electrical charge through a mixture of krypton and fluorine gasses, exciting the krypton ions and producing halogen ions. These two types of ions then chemically combine to form electronically excited but unstable halogen/krypton molecules. These molecules will then split, and in the process emit the photons that act as the basis for lasing. The successful demonstration of this excimer laser generated widespread excitement and stimulated a substantial amount of research in the coming years. High-power excimer lasers, capable of producing light at various wavelengths, were at first used by IBM and others to cut materials used in the manufacture of integrated circuits. Later, they would be widely used in surgery to selectively burn or cut the body.

In the United States, a substantial portion of the laser research carried out during the 1970s was sponsored by the Department of Defense and directed toward the development of military applications. Beginning in the late 1960s, however, the laser industry's focus had begun to shift away from defense research, and the role of the military gradually declined in importance to the field as a whole throughout the 1970s. For one thing, the commercial market grew increasingly important as lasers were used in a variety of industrial applications. Western Electric, for example, announced the first industrial laser system in 1965 with an application designed to pierce holes in diamond dies used for making wire, and over the next ten years a host of other industrial applications was developed. As one observer noted in 1972, "The laser has proved an effective tool in numerous other industrial applications, and increasing numbers of engineers are encountering this new technology in their work. Lasers in industry are being used to measure process parameters and to scribe, drill, evaporate, and weld a wide variety of materials in a wide variety of applications" (Charschan 1972, vii). These included laser-enhanced equipment for use in materials analysis, measuring and leveling applications in construction equipment, control systems designed to accurately position machine tools, and surveying systems.

Fortunately, commercial applications arrived before federal research money began to dry up. Military spending on laser applications continued to increase in the 1970s, but at a slower rate than before, and the passage of the Mansfield amendment in 1969 prohibited U.S. defense agencies from supporting basic research not directly related to their missions. As a result, defense spending as a percentage of dollar value spent on lasers fell from 63.4 percent in 1969 to 55 percent in 1971. At the same time, public opposition to the war in Vietnam forced many scientists and engineers in the field to seriously consider the implications of their work. One university researcher remembered that "[we] were forced to explain the rationale [for our research]. 'Why did you do it? Why did you make those choices?' It was impossible to go through that era without really giving some thought and scrutiny to what you were doing" (quoted in Bromberg 1991, 209). As a result of these changes, the emphasis on both basic research and military applications was reduced and efforts were instead increasingly directed toward commercial and civilian projects that held the promise of either short-term results or of providing substantial benefits to the public as a whole.

As Joan Lisa Bromberg has argued in her book *The Laser in America* (1991), the development of laser technology for use in the separation of uranium isotopes by AVCO Everett Research Laboratories (AERL) is a useful illustration of this shift. The laser separation of uranium isotopes was first developed in 1963 by the French researchers Jean Robieux and Jean-Michel Auclair. Robieux and Auclair used an infrared laser to separate uranium hexafluoride into two isotopes, $U^{235}F_6$ and $U^{238}F_6$. In the second half of the 1960s, as more sophisticated lasers were developed, researchers proposed a wide variety of methods for isotope separation, and in 1969 AERL began a major effort to develop a process for enriching uranium based on these ideas. AERL had been heavily involved in defense research since its founding in 1955, but in the late 1960s they began to explore civilian projects. The director of AERL, Arthur Kantrowitz, later recalled that "we saw the DoD research budget going down, or at least not growing, and we, together with I guess the whole scientific community, got kind of tired of defense work. . . . [It was] the spirit of the late sixties, the Vietnam war . . . . We really searched hard for nonmilitary projects" (quoted in Bromberg 1991, 211). The enrichment of uranium through isotope separation seemed to be a commercially promising civilian venture: in 1970, for example, the Federal Power Commission projected that by 1990 the proportion of U.S. electricity generated by nuclear power would increase from 1.4 percent to 49.3 percent, with a corresponding increase in demand for enriched uranium. In 1971, AERL demonstrated laser-based uranium enrichment for the first time and in 1976, working with Exxon, AERL demonstrated a uranium-enrichment process suitable for large-scale production. Other large laser uranium-enrichment programs were started in the early 1970s as well, including government-financed projects at Lawrence Livermore National Laboratory and at Los Alamos National Laboratory. By the late 1970s, however, the commercial market for enriched uranium had collapsed, due in part to environmental concerns over the use of nuclear power. Ironically, the development of enriched uranium continued to play an important role in the development of nuclear technology for the defense department.

LASERS IN COMMUNICATION

The laser had been conceived in the context of communication devices, so it was not surprising to see such applications come to fruition. However, the laser as a communication device had fallen out of favor over the course of the 1960s. The initial development of the laser excited researchers with the promise of a new communication technology. One commentator, looking back from the vantage point of 1968, wrote that "when the first laser was demonstrated, few people were more excited by it than the scientists and engineers working in the field of communications" (Brown 1968, 65). The initial attraction of the laser was its short wavelength, tight focusing, and rapid speed. Short-range, line-of-sight communication links were soon designed using gallium arsenide lasers, and by 1969 these systems were being tested in Vietnam. However, it was quickly discovered that elements in the atmosphere, including water vapor, oxygen, and nitrogen, absorb certain wavelengths of electromagnetic radiation. This meant that entire bands of wavelengths were unable to travel through the atmosphere for long distances. Furthermore, particles in the atmosphere had a tendency to scatter laser light, and atmospheric turbulence caused unpredictable changes in the laser's refractive index. As a result, transmitting information by laser through the atmosphere for distances of more than a few miles soon struck researchers as an unfeasible goal. When combined with the serious technical and financial problems of alternate proposals, such as closed tubes filled with countless glass lenses, the prospect of major communication systems based on laser technology began to seem less and less promising over the course of the 1960s. As early as 1964, for example, one researcher noted that lasers "do not seem to offer any obvious advantages" for use in terrestrial communication systems (Bromberg 1991, 195).

At the end of the decade, however, lasers suddenly jumped to the forefront of communications research as a result of two breakthroughs. The first was in fiber-optic technology. In 1970 Robert D. Maurer and his colleagues at Corning Glass Works developed a glass fiber system that could transmit light. This was itself not remarkable, since Bell Laboratories had already designed such a system. However, Bell's system lost over 400 decibels per kilometer, which made it impractical for use in telephone communications. Maurer's system, on the other hand, lost only 20 decibels per kilometer, a truly remarkable improvement. Around the same time, the development of double heterojunction laser diodes reduced the current density needed to reach the lasing threshold to the range of 1,000–3,000 amperes per square centimeter. In practical terms, that meant that an efficient, semiconductor laser could operate continuously at room temperature for the first time. This was important in making a fiber-optic system commercially practical. The effect of these simultaneous developments on the field of communication technology was, so to speak, electric: as one overview of the topic recently noted, "The simultaneous availability of a compact optical source and a low-loss optical fiber led to a worldwide effort for developing fiber-optic communication systems" (Agrawal 1992, 4). The first commercial fiber-optic communication systems were begun in 1976, transmitting digitized voice signals at 45 megabits per second over experimental routes a few kilometers long. Technological advances continued throughout the decade, making fiber-optics increasingly attractive. In 1976, silicon fibers were demonstrated that lost only 0.5 decibel/kilometer, and by 1980 commercial fiber-optic systems using InGaAsP laser diodes that operated at 565 megabits per second over tens of kilometers were being built. By the late 1980s, the lasing threshold had been reduced to just 200 amperes per square centimeter and transmission rates increased to 1.3 gigabits/second.

PHOTOVOLTAICS

One of the most promising sources of electrical power generation in the 1970s was photovoltaics or solar cells. The fact that certain materials produce an electrical current when exposed to light was first discovered in 1839 by the French scientist Edmond Bequerel. The effect remained a curiosity in Becquerel's time, while others discovered additional photoelectric materials. For example, in 1873 Willoughby Smith discovered the photoconductivity of the semiconductor selenium, and four years later the first selenium photocells were manufactured. Selenium was not an efficient way to harness the sun's energy, but selenium cells were commonly in use for applications such as automatic door openers by the early twentieth century. The 1954 Bell Laboratories solar cell, discussed in Chapter 1, had an efficiency of about 6 percent, which was approximately fifteen times higher than the best previous solar energy converter and high enough to suggest the feasibility of photovoltaics for power generation.

In 1955 the U.S. Signal Corps was assigned the task of providing power supplies for the first U.S. satellites. Photovoltaic cells appeared to be the ideal power source for space technology due to their lack of moving parts and light weight. In 1958, the first photovoltaic-powered satellite was launched into orbit: *Vanguard I* had a small photovoltaic power system that was used to power a backup transmitter. It remained in operation for eight years, and its success led to the use of photovoltaic technology in various satellites and spacecraft throughout the 1950s and 1960s. At the same time, research also continued in the private sector: in 1959, for example, Hoffman Electronics began to manufacture solar cells with 10 percent efficiency. A year later, this efficiency rate jumped to 14 percent, but then for years there was little progress.

The 1973–1974 oil crisis dramatically increased photovoltaic research and development. By 1978, a cell made at Bell Laboratories achieved 23 percent efficiency, but it was not until 1985 that a team of researchers in Australia were able to duplicate their success. Nonetheless, 10 percent efficiency cells were manufactured in large numbers by the late 1970s, with about 10,000 square meters of such cells being sold by 1978. However, with the temporary end of the energy crisis, funding for such research dried up.

THERMIONIC ENERGY CONVERTERS

A second major area of "alternative energy" research in the 1970s was in the form of thermionic energy converters—devices that transform heat into electricity using heat to release electrons from a hot emitter and then collecting them as an electron current on a cooler electrode. The hot electrode, or emitter, is separated from a cooler electrode, or collector, by an insulating seal. The electrodes are enclosed within a hermetically sealed container, usually filled with an ionized gas such as cesium vapor. Thermionic energy conversion is based on Thomas Edison's discovery that a current seemed to be spontaneously generated when the two electrodes in a light bulb were kept at different temperatures. In 1915 W. Schlichter recognized this as a way to convert heat into electricity, and by the early 1930s the American chemist Irving Langmuir advanced the theoretical understanding of thermionic conversion. Interest continued sporadically into the 1940s, primarily in the United States and the Soviet Union, and in the 1950s several independent groups achieved encouraging conversion efficiency rates of 5 percent to 10 percent. George N. Hatsopoulos and Joseph Kaye of MIT built one of the better known converters in 1958, and work was also pursued in Germany, Sweden, France, and Holland in the 1960s.

The heyday of thermionics followed the Arab Oil Embargo. Suddenly alternative energy sources were in vogue, and governments around the world poured money into research to improve them. Interestingly, thermionics and several other of the most promising alternative energy systems including photovoltaics had already been investigated by the U.S. and Soviet space programs, because spacecraft needed ways to generate their own electricity. By 1977 the Soviet Union had demonstrated the feasibility of thermionic converters in space technology through its Thermionic Power from the Active Zone (TOPAZ) program. Their idea was to use radioactive material to generate heat, which would then convert directly to electricity to power a space craft. The United States had also pursued a similar strategy, but terminated the program in 1973 to focus on fuel cells and solar panels. By the early 1980s, however, the U.S. government again began funding thermionic research as part of the Strategic Defense Initiative ("Star Wars") program. Then, with the demise of that program, the government lost interest before any practical thermionic converter had been built in the United States. The Soviet Union in the 1980s experimentally launched a spacecraft to test its 5 kW, nuclear-powered thermionic generator, but with the dissolution of the USSR this research was discontinued.

5

The Triumph of
Microelectronics

◆

THE END OF THE COLD WAR

Following the end of the Vietnam War and the signing of the SALT missile agreements in the 1970s, the pressures of the Cold War had begun to fade, and with them the need for the better, faster, stealthier long-range weapons systems that had sustained much electron device research. But the small wars that broke out in Africa, Central and South America, Asia, and the Middle East would periodically revive cold war tensions. Coupled with these political factors were technological initiatives that, intentionally or unintentionally, contributed to the revival of the Cold War in the 1980s. Chief among these technology-related issues was the "proliferation" of nuclear weapons and the missile systems to deliver them. Improvements in Soviet and Chinese ICBM missiles prompted President Ronald Reagan (elected in 1980) to announce an ambitious plan for a space-based missile defense system. Not merely a network of early warning satellites, this Star Wars system would also boast powerful laser weapons that would attack incoming missiles in the air. Reagan stepped up research funding for the electronic systems necessary to accomplish these goals, resulting in a significant boon for engineering employment. Then the boom times for engineers brought on by Star Wars suddenly ended in the late 1980s, when the Soviet Union crumbled. The West boastfully announced that it had won the Cold War, though the legacy of nearly a half-century of Soviet-U.S. rivalry did not disappear overnight. Defense contractors argued for continuing support for advanced electronics research, turning from systems designed to counter a monolithic enemy to systems designed for military operations that resembled international police actions. The focus of military technology began to shift from highly destructive nuclear weapons and missile detection systems, tailored mainly to the needs of deterrence, to conventional weapons with pinpoint accuracy, supported by more elaborate networks of global electronic surveillance. This new era in military technology offered new opportunities for research and development.

SPACE ELECTRONICS COMES OF AGE

After finally terminating its Apollo manned interplanetary program in the early 1970s, NASA turned instead to a space station, a reusable space shuttle, and numerous interplanetary or deep-space probes.

The first manned mission of the Space Transportation System, the official name of the space shuttle, was the launch of the *Columbia* on April 12, 1981. The shuttle project marked a significant change in focus for NASA, as it transitioned from operating a small number of purely scientific missions to a large number of more routine launches, often for paying clients. Many of the payloads delivered by the shuttle since its inception have been military satellites, but many missions have consisted of civilian launches or in-orbit scientific experiments.

However, the American/Soviet monopoly in space was soon broken by the European Space Agency, which launched its first Ariane unmanned rocket in June 1981. Designed without the space race pretensions of its competitors, the Ariane was a small, relatively light, and unmanned workhorse for delivering payloads into orbit. Satellite communication systems became a major engineering focus in the 1980s as the shuttle and Ariane reduced the cost of launches. Satellite communication, which had existed in a very limited way since the 1950s, thoroughly transformed electronic communication networks, especially in countries that had not kept pace with the land-based networks built in the United States and Europe. These countries could skip the landline stage of development entirely, using satellites for telephony, radio, television, and computer networking.

Space exploration, particularly unmanned, deep-space missions, also drove space imaging and communication technologies further during the 1980s. In 1983 the Soviet craft *Venera 15* transmitted to Earth the first high-resolution images of the Venus polar area, and compiled a thermal map of most of the northern hemisphere. That same year, the Infrared Astronomical Satellite located several new comets, asteroids, and galaxies, and a dust ring around the star Vega. In 1984, the Soviet-International Vega 1 & 2 sent probes into Venus's atmosphere before continuing to Halley's Comet.

Just as prowess in device manufacturing was shifting both to Europe and to the East, so too was the implementation of advanced electronic systems in space. A telling development was the launch by Japan's Institute of Space and Aeronautical Science of the Sakigake probe in 1985, which rendezvoused with Halley's Comet. Japan, England, France, and Germany were now capable of reaching out into space.

In the space imaging field, probes began to provide unprecedented visual and other data about astronomical objects. The 1989 Galileo mission, for example, transmitted infrared images of Venus and visual images of the asteroid Ida before continuing to Jupiter.

Space-based electronics systems were not always success stories. In January 1986, the space shuttle *Challenger* exploded shortly after liftoff. The failure was blamed on a defective O-ring and "bad management" at NASA. The shuttle program would continue to operate without major incidents for another seventeen years until, on February 1, 2003, when the *Columbia* suffered a mechanical failure and crashed. Satellites also became notorious for postlaunch failures of various kinds, although the shuttle program allowed some of them to be repaired in space. The most famous case of a space repair followed the launch of the Edwin P. Hubble Space Telescope. Only after the telescope was in space did engineers discover a flaw in its design. A new CCD-based camera, designed to correct the flawed optics of the Hubble's large mirror, was installed later by a shuttle crew to correct the problem.

LEGACY OF COLD WAR INNOVATIONS

The 1980s and 1990s saw many of the innovations of the previous three decades becoming entrenched in daily life. Radar, for example, had become so instrumental in the economy that an air traffic controller's strike in the United States in 1981 spurred a direct presidential action to avoid a calamitous shutdown of commercial aircraft travel. Earlier inventions were also spreading beyond their initial markets, sometimes following the development of ways to mass-produce key electronic components such as lasers and microwave tubes. The application of microwave technology, for example, was almost entirely limited to radar, space communication, and long-distance telephony until the 1980s. A specialized application of it in the form of police radar had its roots in the 1960s and became a commonplace in the 1970s before spawning a commercially significant counterpart, the radar detector, in the 1980s. Microwave ovens for use in kitchens appeared in the 1960s, but only in the 1980s did they suddenly become an essential kitchen appliance. Cellular telephony, a system also dependent on microwave technology, got a strong start in Europe and Japan and then took the North American market by storm in the late 1980s. In a similar fashion, the commercial applications of lasers spread after about 1980 to include not only military systems but also a wider range of medical applications, consumer products such as videodisc and compact disc audio players, and eventually even mundane objects such as laser pointers.

THE INDUSTRY

Further changes were underway in the electrical and electronics manufacturing industries. The U.S. consumer electronics industries were nearly defunct by 1980. The few remaining television and car radio manufacturers, such as TV manufacturer Curtis Mathes and once-proud radio and communication equipment manufacturers Delco and Motorola, seemed headed for oblivion. Asian device firms began to decimate their American competitors in the 1980s when they began manufacturing computer memory chips for the first time, a strategy that also helped delay the emergence of a strong European chip industry.

A strong wave of "deregulation" crashed on the communications industries in the 1980s. The major event in the United States was the decision in early 1982 by American Telephone and Telegraph to settle what had become a lengthy antitrust lawsuit filed years earlier by the Justice Department. AT&T, long a monopoly, agreed to divest itself of its twenty-two Bell system companies, which operated the local telephone networks in various regions around the country. The Bell system was to be divided into seven "Baby Bells," while AT&T would retain its long-distance and manufacturing businesses. The breakup had dramatic effects. The Baby Bells still had a virtual monopoly on local service, but numerous long-distance providers sprang up immediately to compete with AT&T. Telephone equipment markets from switching to home telephone sets were suddenly invaded by firms from all over the world. An early leader in the long-distance field was MCI, a company that had, in fact, been a small competitor to AT&T in the long-distance business before the 1980s. Using microwave towers, MCI (which once stood for Microwave Communication Incorporated) had constructed a high-volume "trunk" line between Chicago and St. Louis, Missouri. Building on this backbone, the company rapidly built out its network to link most of the metropolitan areas in the United States.

Once deregulation was accomplished, other companies began creating their own networks using microwave or fiber-optic technology. For example, in 1982 the aged telegraph system of the Southern Pacific Railroad, called the Southern Pacific Communications Company, was sold to GTE. GTE built a new network on the extensive Southern Pacific right-of-way, creating a network that would later become the basis of the Sprint Corporation.

The same decision that resulted in the breakup of AT&T into numerous separate firms also led to the independence of AT&T's Bell Telephone Laboratories. Originally owned by AT&T, it was spun off as Lucent Corporation. Western Electric remained a part of AT&T, but found that the market for telephone equipment had changed. From the consumer's viewpoint, the most noticeable difference was that the old style of heavy, simple, extremely sturdy home and office telephones disappeared within a few years, replaced by cheaper and markedly flimsier devices intended to be replaced every few years rather than used for decades.

These years saw a resurgence in several older U.S. electronics firms, such as General Electric, Motorola, and Hewlett-Packard. Jack Welch in 1981 became the CEO of General Electric, an aging firm that dated to Edison's invention of the light bulb. Welch became an icon of the "New Economy" of the 1990s, especially after author Robert Slater's 1998 book, *Jack Welch and the GE Way*. The company increased its revenues from $25 billion to $90 billion by 1998 and revived GE's image as an innovator.

COMPUTERS AND NETWORKS

A major development since about 1980 was the proliferation of computers in homes, offices, and nearly everywhere else. The integrated circuit–based microcomputer inspired innumerable industrial controls and embedded computer applications. Electrical engineers were among the first to experience this transition as manufacturers redesigned their "bench" test instruments to incorporate computer chips. Meanwhile, although few consumers were aware of it, all around them mechanical and electronic systems were being replaced with microprocessor- or microcontroller-based equivalents including cash registers, ATMs, automotive engine-management systems, VCRs, and gas pumps.

BUBBLE MEMORIES

In terms of the device field, one of the turning points that occurred in the period of a few years on either side of 1980 was the complete triumph of semiconductor memories. Of the several technologies touted as replacements for core memory, perhaps the most hyped was the so-called bubble memory. The Bell Laboratories magnetic bubble memory device was one of the most talked-about experimental electronic devices of the 1960s and 1970s—and one of the greatest disappointments for the engineers who worked on it. Bell Laboratories researchers had worked for years with magnetic or electromagnetic data-storage devices of various kinds for telephone switching and computing purposes. They achieved great success, for example, in the design and manufacture of magnetic core memories in the 1950s and 1960s. However, these memories required external vacuum tube (or later transistor) "driver" circuits, and researchers were constantly looking for ways to make them smaller and cheaper.

Materials scientists, some associated with Bell Labs, had developed certain kinds of magnetic materials in which individual packets or domains of magnetism could be moved around in certain materials, under the influence of applied electric fields. Transistor pioneer William Shockley was brought back to Bell Laboratories on a part-time basis in 1965 and developed an interest in these "domain-wall devices." With his presence, the project began to take on a higher profile. Continued progress in the laboratory led a brash Jack Morton of Bell Labs to announce in 1969 that magnetic bubble memories (as the devices were now known) with capacities of millions of bits would soon be possible.

A crucial breakthrough was the discovery by Andrew H. Bobeck that in certain types of garnet crystals, the magnetic domains formed were round and could be moved around in the crystal easily. Bell Labs researchers were aware of progress in the field of integrated circuit memories and decided that the bubble memory would be a competitor to the magnetic disks and drums then being used in computers and similar applications. They turned their attention to this field, developing a range of serial storage systems (which move large amounts of information into and out of memory in a stream). However, designers of magnetic storage devices did not give up easily, and as the storage density of magnetic disks rose and costs fell, the bubble memory began to look like a white elephant.

AT&T did make an effort to commercialize the bubble memory within the Bell system. The first product to use the new device was announced in 1976, a voice message announcer with a half-megabit of bubble memory holding 24 seconds of prerecorded audio. This was an appropriate product for the telephone industry, where prerecorded announcements were becoming more and more common, and the memory device replaced unreliable magnetic tape–based systems. Bubble memories were also potentially more reliable than the tape or disc memories used in harsh-environment military systems. The publicity generated by AT&T for this new product was substantial, and for a time in the late 1970s it looked as though the bubble would serve niche markets alongside silicon memory technologies then being developed.

However, that market crumbled soon afterward. Andrew Bobeck, one of the leaders of AT&T's bubble memory project, blamed it on technical and economic problems, noting that increasing the storage density of the devices led to problems with impurities in the material and led to lengthy quality-control procedures that made the memories more expensive. With enthusiasm flagging within the organization that had acted as its chief booster, it was unlikely that development would continue at Bell Labs, and indeed the bubble memory was gone by the 1980s.

The Last of Bill Shockley

In the mid-1950s, William Shockley left Bell Laboratories to set up his own transistor firm, bringing with him several of the best and brightest engineers in the industry. By 1956, he had been named co-recipient with John Bardeen and William Brattain of the Nobel Prize for Physics, and was riding high. His employees, however, were becoming less and less enchanted with his mercurial personality, which had also been a problem at Bell Laboratories. Prone to arrogance and angry outbursts, Shockley was rarely an easy person to work with. His abrasiveness was evident to new employees almost from the moment they were first interviewed (he was notorious for giving interviewees special "intelligence" tests), but the allure of working for the famous Bill Shockley was strong.

Not strong enough, apparently, because after just a year of operation, most of the top researchers at Shockley Semiconductor were ready to resign. Eight of them, called "the traitorous eight" by Shockley, did so in September 1957. They left to form the semiconductor division of Fairchild Camera and Instrument, which grew to be one of the most important semiconductor firms of the 1960s.

Meanwhile, Shockley accepted an invitation to become a professor at Stanford University in 1963. While he continued his semiconductor research, he also began developing theories of intelligence and breeding. In 1965, he wrote an article for *U.S. News & World Report* called "Is Quality of U.S. Population Declining?" The article claimed, among other things, that social welfare programs were overwhelming the natural tendency of the weak and feebleminded to be weeded out of a population through disease or accidental death. He singled out the African-American population as mentally inferior to whites, and noted with alarm that their numbers were increasing at a faster rate than whites. All of Shockley's claims had been anticipated by the nineteenth-century science of eugenics, which had been a major influence on twentieth-century racist theories (including the Nazi justification for the Holocaust). The growing strength of the Civil Rights movement had no effect on Shockley's views, and his friends and Stanford colleagues came to shun him. Shockley held on to his unpopular views to the very end, dying of prostate cancer in 1989. His writings and interviews on the subject of genetics were subsequently edited and published by a private press in 1991 as *Shockley on Eugenics and Race*.

PERSONAL COMPUTERS AND THE INTERNET

The early 1980s saw the remarkable increase in sales of personal computers to ordinary people. Through the combined efforts of hardware engineers and software designers, the computer was made smaller, cheaper, and more useful. Millions of people began to buy home computers from Commodore, Osborne, Compac, Amiga, Texas Instruments, and Apple. A memorable moment in the microcomputer field was the introduction in early 1981 of the IBM Personal Computer, or PC, followed in 1984 by the much-publicized introduction of the second-generation Apple computer called the Macintosh. New software, suitable for use by ordinary workers or consumers, transformed the computer into a multipurpose appliance for record keeping, publishing, and entertainment. As unlikely as the personal computer's transformation was, it paled in comparison to the transformation of the Internet into a popular communications medium. Conceived as a research tool and used in its early years almost solely by computer scientists and engineers, the network that became the Internet spread quickly in the 1980s as personal computer users gained access to it, and it began to have a major social impact in the 1990s. The design, manufacture, and use of PCs and networking equipment were bonanzas for engineers that could not have come at a more opportune time, given the many cutbacks in post–Cold War military research and development. The modest beginnings of companies like Apple Computer were legendary by the 1990s, when the second wave of computing technology crested. New firms emerged to provide innovative new computer "peripherals," computer software, or computer services. One of the largest was formed in 1982 as the Control Video Corporation, an online video game company. It later was called Quantum Computer Services, and provided a private online communication service for the Apple and IBM communities. The success of providing online services to personal computer users resulted in rapid growth and another name change: Quantum Computer became the famous America Online (AOL) in 1989, and came to dominate the Internet service provider (ISP) business in the United States in the 1990s.

The lowered cost and continued miniaturization of electronics led to all sorts of other unexpected outcomes, not all necessarily good. Ronald Reagan authorized the CIA to conduct surveillance activities within the United States in 1981, an event that seemed to harken the arrival of a "surveillance society." Cameras were appearing everywhere in public and private spaces, made possible by the emergence of new electronics technologies such as the VCR and low-cost CCD cameras. With the flowering of the Internet in the late 1980s, and especially after the building of the World Wide Web in the 1990s, electronic spying took on a new form, as images from surveillance cameras began to be transmitted across the Web. By the early twenty-first century, according to one source, there were over 1,200 surveillance cameras per square mile in Manhattan, where many residents had their faces captured on camera hundreds of times per day. It was not clear where these and other microelectronics developments of the last two decades of the century were taking society, nor who was in control of them, but it was clear that in the last decades of the twentieth century, life had become saturated with electronics.

THE CONTINUED USE OF VACUUM TUBES

The vacuum tube, a venerable technology that seemed headed for obsolescence back in the 1950s, staged something of a return in the 1980s and afterward. Certain types of vacuum tubes had retained their commercial viability in the 1960s and 1970s, particularly high-power microwave tubes used in telephony, military, and space communication. In some cases, such as AT&T's long-distance microwave telephone system, vacuum tube amplifiers remained in use simply because it was more expensive to redesign or replace the system than it was to continue to use tubes. Microwave tube amplifiers remained in use in AT&T's Transcontinental Telephone Radio Relay Network until 1982, when they were replaced with amplifiers using transistors.

MAGNETRONS AND KLYSTRONS

Two tubes invented in the 1930s were still being improved and used as microwave generators half a century later. Beginning in the 1980s, miniaturized Magnetrons with better performance and enhanced stability emerged, and better designs, including the use of samarium cobalt magnets, improved their power density (a term used to describe the maximum current that can flow through the device) by a factor of more than twenty as compared to comparable Magnetrons of the 1970s. The reliability, compact size, and relatively low cost of the Magnetron led to its continued use in the home microwave oven, sales of which grew explosively in the 1980s. At the end of the twentieth century, tens of millions of Magnetrons for microwave ovens were sold each year. Klystrons used as efficient, high-gain amplifiers saw some improvement, and were still widely used in applications such as UHF television transmitters and other communication systems. Because they can handle extremely high levels of power, Klystrons were also used in scientific applications such as particle accelerators, where microwave power of up to hundreds of megawatts was needed.

Ultimately, though, by century's end it began to look as if the days of the tube in microwave applications were numbered. One by one, semiconductor devices were devouring the remaining applications. Further, new applications for microwaves were appearing in which vacuum tubes simply could not compete. In 1960, Carver Mead (an engineer more famous for his later contributions to the design of VLSI chips) constructed what was apparently the first GaAs metal-semiconductor (or metal-Schottky) field-effect transistor (MESFET). This device, a variation on the MOSFET, allowed operation at higher frequencies than a traditional MOSFET, and held out the promise of inexpensive, highly integrated microwave circuits. However, it was more difficult to fabricate than a regular MOSFET, and remained in the laboratory for over a decade. It reemerged in the late 1980s, just as interest in inexpensive microwave devices was beginning to rise. By this time, microwave communication was no longer limited to satellites and long-distance telephony. Now, microwaves were coming into wide use in cellular telephones and other consumer devices.

GYRO-DEVICES AND THE
FREE-ELECTRON LASER

Another important trend has been the commercialization of the gyro-oscillator/amplifier and the free-electron laser (FEL, a type of laser that depends on vacuum tube technology), both of which were invented in the 1960s. Improvements in gyro devices led to the practical utilization of scientific instruments using the 1–10-millimeter wavelengths. A major driver of these improvements was the desire to use microwave heating in advanced research on nuclear fusion. Such "plasma" heating was also the driving force behind improvements in the free-electron laser. Today's FELs generate power at submillimeter wavelengths in the infrared range, and it is anticipated that the range of these tunable devices can be extended to the X-ray band.

TRAVELING-WAVE TUBES

Traveling-wave tubes (TWTs) made significant strides in the 1980s through changes in design. For example, the lifespan of one class of TWTs increased from about 9.5 to 15.8 years while their maximum power output increased from 20 to 110 watts, and their weight decreased from about 1.2 to 0.86 kg while the prices of these tubes dropped significantly. More recently, hybrid vacuum-semiconductor microwave power modules have emerged, which combine the best features of both TWTs and semiconductor electronics.

Four companies dominated in the manufacture of these components by the 1990s: Hughes Electron Dynamics Division in the United States, Thompson Tubes Electroniques in France, AEG Microelectronics in Germany, and TMD Technologies Limited in the United Kingdom. France, Russia, and the United States remained the centers of TWT research.

Many of these microwave transmitting and receiving tubes would be barely recognizable as vacuum tubes to someone familiar only with the technology of the 1950s, but they did retain the technological basis of the original Audion tube. Further, much of their continued success owed a great deal to the proliferation of satellite communications systems beginning in the 1980s. Direct Broadcast Satellite television service, for example, was launched in 1983, giving space-based microwave tubes (not to mention other space electronics systems) a new boost. The Satellite Television Corporation, a subsidiary of Comsat Corp., was given a contract to build a satellite for this system using a 200-watt traveling wave tube amplifier.

TUBES FOR AUDIO

At the close of the 1970s, there were still a few manufacturers of vacuum tube–based audio equipment, manufacturing extremely expensive amplifiers for the "audiophile" market. They depended heavily on old stocks of tubes that were by the 1980s getting increasingly hard to find. In the United States, Japan, and Europe, the manufacture of tubes suitable for audio had virtually ceased by 1988, when the last Western Electric type 300B triode, a design first offered in 1938, ceased production. By that time one of the major customers for the tube was a Japanese consumer electronics company called Laser, which was making small batches of amplifiers. In 1990, the Kansas City Western Electric plant sold the last of its inventory.

Yet by that time the vacuum tube amplifier for home audio was enjoying something of a comeback, and dozens of small companies began offering designs, many of them harkening back to the famous Williamson high-fidelity amplifier circuit of the 1940s. With American and European-made tubes becoming scarce, prices shot up. Seemingly out of the blue, a new source of tubes became available. With the dissolution of the Soviet Union in the late 1980s, Western audiophiles became aware of manufacturers in the former Soviet Republics and China who were still making many older tube types. Tubes were still in demand in these countries because older, tube-type military equipment and consumer radios were still in active use there. The consumer audio uses of vacuum tubes actually expanded in the last decade of the twentieth century, as some manufacturers have incorporated them into the designs of new types of equipment such as CD players.

THE LED AS A DISPLAY TECHNOLOGY

The LED, once the basis of an evolving electronic display technology, was pushed aside by other technologies in many applications by 1980. The most significant change in the LED field was the development in the early 1980s of a liquid-phase-epitaxy manufacturing technology, which allowed LEDs to be made from gallium aluminum arsenide (GaAlAs) rather than silicon or GaAs. This was announced by National Semiconductor Optoelectronics (formerly Xciton) in 1982. GaAlAs provided dramatically superior performance over older LED technology, with ten times greater brightness due to increased efficiency. These new LEDs were three to twenty times as bright as comparable incandescent lamps, and in the late 1980s clusters of LEDs began to replace incandescent bulbs in outdoor applications such as automobile and truck brake lights, traffic signals, airport runway lights, and advertising signs. GaAlAs LEDs required only a low voltage to operate, making them more cost-effective than incandescents in many applications. They were soon designed into systems such as bar code scanners, fiber-optic systems, medical equipment, and other applications. Infrared emitters made possible by GaAlAs LEDs, for example, became common in TV controls, timers for disk drives, end-of-tape indicators, and optical switches.

Unfortunately, the new LEDS still faced two serious drawbacks. Firstly, GaAlAs could only be used to produce red LEDs. Yellow, green and, orange LEDs saw only minor improvements during this time stemming from developments in optics technology and crystal growth techniques. Secondly, the light output of GaAlAs LEDs tends to decrease by as much as 50 percent after only 50,000–70,000 hours of operation. This was especially a problem in hot, humid environments. LED designers attempted to address these problems in a number of ways, the most important of which was the use of laser diode technology to develop Indium Gallium Aluminum Phosphide (InGaAlP) LEDs. InGaAlP LEDs, first introduced for lighted signs and other outdoor applications around 1990, are characterized by very high brightness and reliability. Because certain aspects of the design of an InGaAlP permits adjusting the output color, this type of LED provides more flexibility than other types. As a result, red, yellow, orange, and green LEDs can all be produced using the same basic technology. Additionally, light output degradation is significantly improved over GaAlAs LEDs, even in hot and humid environments.

By the end of the 1980s, the development of commercially feasible organic LEDS (OLEDs) began to seem like an attainable goal. There are certain organic materials called conjugated polymers that behave somewhat like semiconductors and can be made into diodes. Their first applications included their use in copy machines, where they played a part in detecting and reproducing images on paper. Later, Ching Tang and Steve Van Slyke of Kodak found a way to use them as light emitters. Because of their brightness and ease of fabrication, many predicted that OLEDs would replace ordinary semiconductor LEDs. The organic materials also made it relatively easy to construct high-density arrays of OLEDs, leading the Pioneer Corporation to introduced the first OLED computer display.

LCDS

Liquid crystal devices (LCDs), which had been commercialized as competitors to simple LED numerical displays in the 1970s, by the later 1980s would also be used for video and computer display purposes. It was this commercial application that propelled LCD and related technologies to preeminence in the display field. Toshiba was apparently the first to offer a relatively large LCD display in a personal computing product in 1982, when it incorporated one into its stand-alone word processor.

Since an ordinary LCD reflects light but does not generate it, practical LCD displays usually had to be back-lit, and this made the design of battery-operated devices more difficult. LCDs, still capable of just black-and-white displays in the early 1980s, were helped along by the development of so-called active matrix technology, in which thin-film transistor circuits added to the LCD matrix acted as precision modulators of the un-twisting of the liquid crystal structure. Now, instead of a pixel being either "on" or "off," it was possible to give the image the subtle variations of grayscale, such as those available on a monochrome television display. The active matrix design also makes it possible to address each pixel individually rather than scanning one at a time, and this aids in making rapid changes to the image (as in television, games, or other moving images). The development of active-matrix technology led to the development of relatively large LCD arrays, thus making applications such as flat-panel displays for computers and televisions possible. Of course, "large" is relative—the first LCD televisions introduced in 1983 and 1984 by Sony and Seiko had screens with diagonals of just a few inches, increasing to 6–10 inches by 1985.

Color LCD displays for computer and television purposes were first demonstrated in 1988 by Sharp Corporation. In such color displays, each pixel is divided into three subpixels, each with its own LCD, which is seen through a red, green, or blue filter. Several companies demonstrated color, thin-film-transistor, liquid crystal displays with screen diagonals of up to 14 inches. While this was a rapid increase, it is worth remembering that 14-inch televisions (using CRTs) had become available before 1950. While the CRT's place in television systems remained safe for the time being, LCD displays quickly took the portable market.

There were numerous improvements to the basic LCD technology in subsequent years, as the displays began to be incorporated into cellular telephones, handheld computers, games, and other devices. By 1995 around 10 million LCD arrays were produced for personal computers, mainly portable "notebook" types. This number doubled by 2000.

THE CCD COMES OF AGE

The inventors of the CCD at Bell Laboratories had imagined that the device might best be utilized as a high-speed memory device or shift register, competitive with another AT&T product, the bubble memory. As it turned out, the capabilities of both the bubble memory and the CCD memory were unexpectedly surpassed by ongoing developments in semiconductor memories. Both were relegated to the dustbin, but the inherent flexibility of the CCD gave engineers opportunities to apply it elsewhere. As a computer chip, the CCD enjoyed little commercial success, but it made a more dramatic impact as a replacement for photographic film and video camera tubes.

The introduction of the CCD represented a major change in the field of space-based imaging systems. High-density, CCD linear array strips were used beginning in the 1980s to view a strip of surface area (such as the surface of the Earth), and the output of those strips were electronically joined to form a larger image. The first system to use such a "pushbroom" scanning arrangement was a German-designed satellite. Later CCD imaging devices, such as the famous Hubble telescope, used larger arrays to capture larger images instantaneously. CCD imaging technologies were much more sensitive than comparable photographic or Vidicon (vacuum tube) systems, and subsequently the resolution available from space-based cameras rose very quickly. The most sophisticated CCD imaging satellites, capable of resolving objects on the ground as small as a meter or two, were launched by the military and used for controversial intelligence-gathering purposes. Mass production of the CCD eventually led to its incorporation in consumer video and still cameras. By the 1990s, the CCD was the commercially dominant form of the electronic imaging device.

SOLID-STATE SENSORS: THE ADVENT OF MEMS

A significant development in the 1980s was the emergence of many new types of solid-state sensors for detecting light, pressure, temperature, mechanical position, or other features of the "real world." So many new technologies were going digital that engineers were looking for ways to better integrate traditional analog sensors with digital equipment. Some solid-state sensors, such as those based on piezoelectric devices, had existed for decades, but there were only a few of these. As the applications for digital electronics grew, and especially as they began to include consumer items, the need for inexpensive sensor technology grew. Early examples of the new generation of technology were the Motorola and Honeywell Corporations' solid-state pressure transducers for automotive engine control systems, introduced in 1980. They featured a direct voltage output that varied in response to pressure, and were manufactured using existing IC fabrication techniques.

A new class of integrated circuit, often used as a sensor, was micro-electromechanical systems (MEMS), first proposed in the 1960s but not widely commercialized until the 1980s. The idea behind MEMS devices was to use integrated circuit fabrication techniques to make tiny mechanical devices, some with moving parts, which could be linked to electronics made on the same chip. One of the first practical applications was the micromachined nozzle assembly used in the cartridges of inkjet printers. In 1982, automotive airbag systems (which had been proposed in the 1950s) were reintroduced using MEMS sensors to detect a crash. The Analog Devices Corporation elaborated this idea, producing an "accelerometer" for airbag systems in 1991, where the mechanical and electronic portions were integrated on the same chip. The company later introduced a gyroscope on a chip that was capable, for example, of working with an automobile's Global Positioning System (GPS) to create more accurate maps and directions for drivers. MEMS devices, because they are so tiny, were also beginning to be used inside the human body for a variety of purposes by the 1990s. Proposed applications included the monitoring of control of failed heart valves, the operation of tiny insulin pumps, and numerous other uses.

INTEGRATED CIRCUITS EVERYWHERE

The convergence of the integrated circuit (IC) and the computer with communications technologies, already underway in earlier decades, began to transform virtually every segment of society by 1980. A considerable amount of research went into production processes for existing technologies, or improvements (sometimes quite dramatic) to the design of such now-standard items as computer memory and microprocessor chips. Internationally, Asian and European firms began to catch up and sometimes surpass the pioneering American research institutions, particularly in bringing devices to market. A crisis in American memory chip production led to a drastic rethinking of the whole chip industry, while the later collapse of the Soviet Union resulted in a wholesale scaling back of device research there. Defense funding still supported a considerable amount of new device research, but the end of the Cold War meant that research efforts were moving in new directions.

VLSI

A new buzzword in the IC field in the late 1970s had been VLSI—very large-scale integration. Struck by the rapid increase in the number of elements that could be incorporated into memory ICs, engineers and marketers sought a new way to describe the exponential increases in the number of individual elements that could be crammed onto a single chip. "Large-scale integration" seemed to suffice for a while, but the almost daily advances in technology led to its rapid obsolescence. In the 1970s, LSI described chips with at least 1,000 active elements, while VLSI referred to chips with tens of thousands of active elements. By the early 1980s, when the first VLSI chips came to the market, the term was already a part of the engineering lingo.

In the 1980s, engineers increased performance of MOS chips by replacing the aluminum gates of the transistors on these chips with polysilicon. Polysilicon gates made it practical to employ both n-channel and p-channel MOS transistors on the same IC—and thus one could design a complementary MOS (CMOS) chip. The principal advantage of CMOS was its lower power consumption, and the use of polysilicon gates actually simplified fabrication and allowed for smaller device sizes. It was the CMOS type of chip that dominated the VLSI field by 2000.

Texas Instruments introduced a 64-KB RAM chip in 1979 that was widely considered the harbinger of the VLSI wave. Development of the chip had been partly sponsored by the military, but it was unclear whether other markets existed for such high-density ICs. However, the drive to integrate continued, and as prices came down, markets opened up. 1981 saw 64-KB memories come to the market from at least three manufacturers.

Meanwhile, the older technology of LSI was being used to create chip sets for mainframe computers and the new supercomputers, such as the three-chip mainframe introduced by Intel in 1982 or the supercomputers on chips offered by Control Data and Burroughs. Bell Laboratories was also doing more with less and showing how integrated circuits could be used for complex systems other than just computers; its largest chip was a time slot assigner for the T-1 digital transmission line system, distributing any of twenty-four incoming signal streams to any of 256 outgoing lines.

By 1983, several companies had already introduced 256-KB dynamic RAMs, though production of these was just getting underway when IBM announced its 512-KB RAM memory the next year. At this point, the well-established NMOS construction technique was being replaced by CMOS for many new memory chips, and Intel was in the process of designing new CMOS microprocessors too. By about 1990, CMOS gate-array ICs commonly incorporated over 200,000 gates, though chips with around 100,000 gates were still more common. The progress in memory chip fabrication (and the falling price of memory) was so pronounced that by 1986–1987, several companies introduced variations of the EPROM chips intended to replace the dominant storage technology of the personal computer—the magnetic disk drive. The first such "hard cards" included those introduced by Toshiba in 1987. This event was indicative of the rise of nonvolatile "flash" memory chips, devices capable of retaining information even if the power was cut off (as it would be in a card removed from the computer). These chips proved to be useful not only in devices like hard cards, but also in office and home appliances (such as facsimile machines) and even credit cards. Nonvolatile flash memories were proposed by Toshiba in 1985 and emerged a few years later. Texas Instruments, for example, brought out a flash EEPROM (electrically erasable programmable ROM) with 256 kilobits in 1989. The EEPROM could be reprogrammed if necessary without the need for an ultraviolet light or removal of the chip. Many predicted that the magnetic disc would soon be obsolete, but manufacturers had a few tricks left, and began improving the performance of small hard discs. Their efforts made discs competitive with chip-based storage through the end of the century.

Gordon Moore on Moore's Law, 1997

Gordon Moore is a founder of Intel Corporation.

I first observed the doubling of transistor density on a manufactured die every year in 1965, just four years after the first planar integrated circuit was discovered. The press called this "Moore's Law" and the name has stuck. To be honest, I did not expect this law to still be true some 30 years later, but I am now confident that it will be true for another 20 years. By the year 2012, Intel should have the ability to integrate 1 billion transistors onto a production die that will be operating at 10GHz. This could result in a performance of 100,000 MIPS, the same increase over the currently cutting edge Pentium II processor as the Pentium II processor was to the 386! We see no fundamental barriers in our path to Micro 2012, and it's not until the year 2017 that we see the physical limitations of wafer fabrication technology being reached.

Source: Dr. Gordon E. Moore, chairman emeritus, Intel Corporation, "The Continuing Silicon Technology Evolution inside the PC Platform," http://www.intel.com/update/archive/psn/psn10975.pdf.

MANUFACTURING

A good bit of the continued development of VLSI can be attributed to advances in manufacturing rather than breakthroughs in new devices. Ion implantation, developed in the laboratory in the mid–1970s, did not come into wide use until the 1980s. Another technique invented in the 1970s, plasma etching, began to challenge "wet" chemical etching in the processing of chips by the 1980s. There were other important production techniques as well, such as the advent of random (reactive) ion etching (RIE).

The process of photolithography, which had become universal in the manufacture of ICs since their inception, seemed to be reaching its physical limits in the early 1980s, when engineers believed that lines narrower than 1.0 μm would be impossible. Line width, the term used to describe the smallest possible features that can be fabricated on a chip, had decreased from a minimum of about 5 μm in the 1960s to about 1 μm by the late 1970s. Few observers in 1980 believed that optical lithography would persist much past the late 1980s, if such reductions in size were going to be continued. In fact, optical techniques were destined to dominate production through the end of the century. Researchers printed features as small as 0.8 μm in the early 1980s, and this was done in a production setting a few years later. The smallest lines possible were 0.5 μm by the early 1990s, shrinking to just 0.35 μm by 2000. Part of that improvement came from the use of single-wavelength exposure at 436 nm (known as G-line), 405 nm (H-line), or later 365 nm (I-line) or 248 nm (deep-UV) wavelengths. I-line exposure, which depended on the development of improved quartz lenses, proved to be the most widespread technique by the 1990s.

E-BEAMS AND X-RAYS

There were notable efforts to improve other forms of lithography. The technique of electron-beam lithography, which promised to produce small chip features in fine detail, was introduced in commercial form by companies such as Leptron of Murray Hill, New Jersey. The Leptron EBES4 could write up to five 8-inch wafers per hour with feature sizes smaller than 0.4 micrometers. Unfortunately, this was too slow to be cost-competitive with optical lithography, but electron beam techniques were used to create the masks used for making chips by the late 1990s. Bell Laboratories continued to improve the X-ray lithography techniques first announced in the 1960s, but this process remained in the experimental phase through the early 1990s, when IBM finally started making prototype chips using a huge synchrotron-based X-ray lithography machine. Part of the reason why optical lithography remained vital was simply inertia and familiarity, but also its incremental improvements in performance. At the end of the century, engineers were once again talking about fundamental physical limits and the need for a radically new way of making ICs (or perhaps an entirely new kind of device).

MICROPROCESSORS

What finally brought the integrated circuit out of relative obscurity and into the attention of the general public in the 1980s was the microprocessor, that computer on a chip that formed the heart of many automated systems and embedded control devices. Following the introduction of the Intel 4004, the company introduced its more powerful 8008 and 8080 chips. Texas Instruments, Motorola, and Zilog (a new firm cofounded by former Intel designer Federico Faggin) had also entered the marketplace.

When the Apple Computer appeared, it used a variation of the type 6502 microprocessor designed by MOS Technology, a new company formed by former Motorola employees. Chips based on the 6502 would find their way into later Apples as well as the Commodore personal computer. However, in 1981 IBM selected the Intel 8088 (introduced in 1978) for its new Personal Computer. The rapid proliferation of IBM and "clone" PCs fueled work toward improvements at Intel, resulting in the subsequent 80286 (1982) , 80306 (1985), and 80486 (1989) devices. Competing products introduced in the 1980s included those by IBM, Zilog, Motorola, Digital, Sun, Hewlett-Packard, and dozens of other (including Soviet and Western European designs). Significantly, the Macintosh computer introduced in 1984 used a Motorola 68000 series microprocessor (a type first offered in 1979), and the strong market for these computers in the 1980s and early 1990s spurred a series of improved models, including the 68020 (1984), 68030 (1987), and 68040 (1990).

Up to about that time, "microprocessor" had not yet fully entered the lexicon. Then in 1991, Intel launched its "Intel Inside" advertising campaign. Working with manufacturers of personal computers, Intel "cobranded" the computers so that consumers would know that they used Intel microprocessors. Whereas before, awareness of the innards of a computer was rarely a concern to most consumers, suddenly the brand of microprocessor became as important as, say, one's choice of aftermarket car stereo. The Intel Inside campaign was enormously successful, especially

when it was combined with the launch of Intel's "Pentium" microprocessor in 1993. No microprocessor product attained the Pentium's level of
public recognition through the end of the century.

A major divergence in microprocessor design was the introduction of
reduced instruction set computer (RISC) chips in the 1980s. Up to that
time, personal computers had employed what was retroactively seen as
"complex" instruction sets. By reducing the number of machine language
instructions, much complexity on the chip could be eliminated, leading to
more efficient operation. More complex but rarely used operations were
then written into software rather than designed into the chip. The most
notable early example of an RISC chip was the Motorola PowerPC introduced in 1993.

ASICS

The early 1980s also saw more engineers designing complex systems using
a combination of standardized chips (such as microprocessors and memory)
supplemented by custom-made ICs. These application-specific integrated
circuits (ASICs), tailored to a particular system, combined the functions of
a number of standard chips into one device, reducing the overall number of
ICs required for the system. There had always been application-specific
chips made, but in the 1970s it became more common for manufacturers to
promote standardized chips that sold in large numbers versus customized
chips made for high-end computers, for specialized instruments, or sometimes for the military. As companies began to look for more efficient ways
to make their custom chips, they began to promote certain standardized ap-
proaches to their design. A key component of the story was the introduction some years earlier of computer-aided design technology, or CAD. By
reducing the design process to a set of rules, computers could aid chip designers and substantially reduce the time it took to lay out chip masks. Bell
Laboratories, for example, reduced the cost of the technology it used for
echo canceling on telephone lines from a $10,000 circuit to an inexpensive
ASIC, but this could not have been done as economically without the help
of CAD. Further, manufacturers began to produce semifinished chips with
clusters of standardized circuits on them, such as gate arrays, which could
be interconnected to form complete ICs to a customer's specifications by
completing the manufacturing process. The market for these chips was
growing by 25 percent a year by 1980, and the end seemed nowhere in
sight. It was an important development for American and European chip
manufacturers, beleaguered by Asian competition in the memory field.

As ASIC design matured, most designers turned to assembling pre-designed blocks of circuits, speeding the design process. So, for example, most designs continued to use off-the-shelf microprocessors, PROMs, DRAMs, and SRAMs, but they used custom-designed chips for such applications as voice processing, for the interface between microprocessor and memory, or for products with very small production runs. For example, Sun Microsystems' popular SPARCstation 1, introduced early in 1989, made extensive use of ASICs to achieve better performance at lower cost than its competitors, reduce size, and improve reliability. The ASICs used were typical of those developed for similar projects, and included the integer unit (IU), floating-point unit (FPU) cache controller, memory-management unit (MMU), data buffer, direct memory access (DMA) controller, video controller/data buffer, RAM controller, and clock generator. ASIC design was a well-established field by the 1990s, with the devices often taking over functions that required DRAMs and/or PROMs, EPROMs, or EEPROMS, such as input/output circuits, timing circuits, and any number of other special functions.

DIGITAL SIGNAL PROCESSING

In the field of digital signal processing (DSP), ASIC chips gained such prominence during the 1980s that they merit a separate discussion. Some of the earliest DSP chips appeared in the late 1970s from manufacturers such as AMI, Intel, NEC, and Texas Instruments. As IEEE historian Frederik Nebeker has demonstrated, DSP chips evolved alongside sophisticated algorithms needed to manipulate digital video, audio, and other data. Signal processing was used to artificially generate sounds or images, to reduce "noise" and other unwanted distortion, to enhance audio or image quality, and especially to compress digitized audio or video in order to transmit it more efficiently. Emerging from high-technology communications, aerospace, and military applications of the 1960s and 1970s, IC versions of signal processing circuits quickly made their way into consumer technologies. One of the first was a children's toy, the Speak & Spell by Fisher-Price, which contained an innovative speech-synthesis chip manufactured by Texas Instruments. Two years later, a toy that had been a perennial favorite since the late nineteenth century went solid state, when Fisher-Price introduced the first talking doll incorporating DSP. The chip was made by Precision Monolithics, Inc., of Santa Clara, California. These chips were rapidly incorporated into radar systems, fax machines, camcorders, videotape equipment, satellite imaging systems, medical electronics, and dozens of other emerging technologies. Interest in so-called neural networks (electronic systems capable of simple learning), which had declined in the 1960s and 1970s, revived in the 1980s when it began to be applied to image processing, speech recognition, machine vision, and other applications, and several firms pioneered in manufacturing VLSI neural network chips. Alongside this came the desire for "parallelism" in signal processing, that is, dividing a complex computer task into two or more tasks and using multiple processors to perform it. Texas Instruments introduced the first DSP chip designed for parallel processing in 1991, with six ports for direct interprocessor communication, a six-channel compressor, dual external busses (the data conduits between chips), and a processing speed of 275 million operations per second. A major stimulus to the DSP market occurred as digital fax machines overtook analog predecessors, as personal computers and videogames began to incorporate much more sophisticated color graphics capability, and as modems came into widespread use during and after the mid-1980s.

This period saw a dramatic increase in the market value of DSP chips, rising from perhaps $50 million to $2.2 billion between 1985 and 1995. Many of these were so-called converter ICs (i.e., chips for converting digital to analog data or vice versa), which are central to sensing applications, guidance and control, navigation, and medical systems. The United States and Japan remained world leaders, while Russia (which in Soviet days had been a leader in military DSP) lost its ability to conduct leading-edge research.

THE CONTINUING INFLUENCE OF THE MILITARY

The twenty-year period between 1980 and 2000 saw changes in the relationships between the military and the microelectronics industry. In the United States, the Department of Defense had been an important source of research support and a major customer for many new technologies, from miniaturized vacuum tubes in the 1940s to the transistor after 1947 and the microprocessor in the 1970s. The availability of research money also sped up the development of VLSI chips of the late 1970s and early 1980s. The U.S. Department of Defense, for example, partly funded the development of a Texas Instruments 64-KB dynamic memory chip introduced in 1979.

At about the same time, the U.S. Department of Defense announced a major push for what it called very high-speed integrated circuits, and annual funding for the project was expected to reach $200 million during the next few years. The DoD funded the development of this and other semi-conductor technologies that were widely considered too expensive for the nonmilitary market, including GaAs chips. Invented in the 1960s, GaAs transistors and chips were under development at Rockwell Corporation, Bell Labs, RCA, Hughes Electronics, Westinghouse, TRW, Fujitsu, NEC, and Toshiba, as well as numerous university and military labs. Many believed that GaAs held the promise of very high-frequency operation, which would be useful for microwave integrated circuits and supercomputers.

By 1982, monolithic microwave integrated circuits or MMICs were being tested in military systems. Monolithic GaAs power and low-noise amplifiers were used in phased-array radar, where a switched array of small, flat radar antennas replaced a large, mechanically steerable antenna. Military GaAs chips were also being used experimentally for electronic warfare purposes, to jam communications. MMIC development was also taken up in Europe, where it was believed that a nonmilitary market existed in satellite communication. Within a few years, these chips were entering the commercial sector as the basis of receivers for the new Direct Broadcast Satellite television system.

The DoD and DARPA, the Defense Advanced Research Projects Agency, also inaugurated in the early 1980s an important program to upgrade military electronics and improve reliability and maintenance. The military faced the problem of obtaining spare parts for electronic systems that were, in some cases, up to twenty years old. Various branches of the services attempted to solve this problem by sponsoring the construction of special manufacturing facilities capable of quickly turning out relatively small batches of custom integrated circuits, sensors, and microwave devices. DARPA, better known for its role in creating the Internet, was also intimately involved in the transition of GaAs IC technology from the laboratory to the production line. The agency sponsored the construction of a GaAs IC fabrication line in 1983 as part of the program to upgrade military electronics, making chips to be used in satellites for onboard signal processing. The military's program for very high-speed integrated circuits, which hinged on GaAs technology, began to bear fruit in 1983, when TRW, Hughes Aircraft Co., and IBM announced the first IC product, a matrix switch, and announced upcoming designs for specialized signal processing applications. The military sponsorship of GaAs technology early in its history helped bring the technology to nonmilitary markets as well. Sales of GaAs ICs by 1990 reached a quarter billion dollars per year, and had spread from military systems to commercial high-speed computers and high-frequency communications. Clearly, here was a successful example of a military spinoff of a product once thought too expensive for the commercial market. While GaAs chips have never fulfilled the expectations they generated in the 1960s, they are today widely used in several types of communications systems, including cellular telephony.

DIFFUSION

While it had seemed that the integrated circuit had achieved a great deal of success in the 1970s, there was more to come during the next two decades. This applied not only to the vast array of new electronic devices becoming available, from digital cameras and cellular telephones to handheld computers, but also to products that had in the past been nonelectronic, or even nonelectrical. It was a process of diffusion, where seemingly everything was "going digital."

One of the areas outside computing where the microprocessor began to have an impact was in the field of instrumentation. These test and measurement devices, so dear to the hearts of electrical and computing engineers, very often served as test beds for digital devices that would later appear in consumer products. Instruments in 1980 were already relying on microprocessors for self-testing, diagnostics, and autocalibration features. A Philips oscilloscope manufactured early in the 1980s was one of the first to use charge-transfer (similar to CCD) devices in an onboard analog-to-digital converter. Digital storage oscilloscopes, capable of storing waveforms so that they could be displayed on the screen indefinitely, were offered by Tektronics and Hewlett-Packard, two of the leading U.S. instrument makers. Even simple instruments such as multimeters (handheld voltage/current/resistance testers) were going digital, with LCD displays and microprocessor-based electronics. By middecade, some wondered if these digital instruments were not in fact becoming a form of the personal computer. The *IEEE Spectrum* noted that an oscilloscope in the 1986 Tektronix line boasted three microprocessors, a graphical interface with pop-up menus, a touch-sensitive screen, a keyboard, and sophisticated, built-in software. One high-end Hewlett Packard oscilloscope of the same year even had a color display. This was a prescient observation, for in the 1990s PC-based circuit boards could be added to a desktop computer to give it advanced instrumentation functions, although standalone instruments also remained on the market into the twenty-first century.

One reason for the early appearance of so many device innovations in instruments is that these instruments are the equivalent of machine tools in the metalworking trades: that is, they are machines to make machines. Because of this, instrument makers were compelled to design systems that could exceed the accuracy of the devices they were testing, and hence they had to rely on specialized versions of some of the same technologies being deployed in computers, communication equipment, and other instruments. An example of this was the fiber-optic technology incorporated as early as 1986 into power meters, variable attenuators, and devices used in testing fiber-optic communication equipment.

Electronics also made unexpected gains into transportation technologies at an early date. From the vacuum tube auto radios of the 1930s to the transistorized ignition systems of the 1960s, the automobile had for some time been slowly accumulating more electronics, even before the advent of digital devices. Through the 1970s, particularly in response the U.S. tailpipe emissions standards, automobile engines were increasingly equipped with microprocessor control and monitoring technology. The first such system was probably the Motorola microprocessor used in GM cars in 1975. With even stricter emissions requirements, the coming of a second major energy crisis, and heightened concerns about auto safety in the 1980s, automobile manufacturers began deploying more microprocessors in new car designs to manage engine functions, control suspensions, deploy air bags, and control other systems. Early designs used microprocessor-based engine management units to control a bewildering array of vacuum- or solenoid-driven analog fuel and ignition system controls, leading to engine compartments that were a mass of hoses and wires. The worldwide adoption of electronically controlled fuel-injection systems killed two birds with one stone, simplifying the tasks of manufacturing and servicing auto engines by reducing the number of components, and enhancing the performance and efficiency of the internal combustion motor while simplifying the task of engine management.

The electronification of transportation did not end there. By the mid-1980s, engineers had developed all sorts of microprocessor-based automotive systems ranging from antilock brakes and improved automatic transmission controls to "active" suspensions that adapted a car's handling to driving conditions. These were all longstanding aims of automotive engineers, and all were previously implemented with limited success by mechanical means. In 1990, Analog Devices introduced the first single-chip IC accelerometer for automotive airbag control, the beginning of a string of chip innovations that helped make airbag systems affordable.

Moreover, autos were not the only forms of transportation to have digital controls grafted onto their existing designs. Aircraft had long been heavily dependent on electronics for communication and guidance, and soon "fly by wire" airliners that substituted electronic networks for mechanical or hydraulic flight controls were everywhere in the skies. In San Francisco, engineers also employed electronic controls in the Bay Area

Transit (BART) System, a controversial but ultimately successful new electric traction system. A 1984 update of the BART system placed three redundant microprocessors in each car. Freight train controls had already been taken over by electronics by the mid-1980s, and these were also superceded by microprocessors.

In consumer electronics, where radios and televisions had begun to incorporate analog integrated circuits as early as the 1970s, usually as a cost-cutting measure, now new features were added to these products through the incorporation of microprocessors. RCA, for example, introduced a 19-inch TV with an NMOS microprocessor in its 1979–1980 line, which allowed the user to program the set up to a week in advance. This was intended for use with a VCR, but was not a great commercial success. More successful were microcontrollers used in VCRs (introduced in 1975) themselves; these performed numerous control functions, including making the clock blink. The early 1980s saw Phillips Electronics standardize a simple, inexpensive, two-wire data communication system (called the I2C bus) to allow microprocessors in appliances to communicate with other chips in the same system. Soon the microprocessor was in seemingly everything, from gas pumps to washing machines, remote controls, and "smart cards" providing better controls, or sometimes simply more gimmicks. The still-developing field of home automation was also born about this time, when in 1984 General Electric introduced its "Homeminder" system. Using a television receiver–based interface and microprocessor control, the Homeminder automated lighting and appliance control. Whether these microelectronics devices were employed to save money, improve efficiency, or simply add "bells and whistles," there was no denying the profound electronification of nearly every form of everyday technology by 2000.

Medicine has been another area where electronic devices have moved from a very limited role to a central place. Nuclear magnetic resonance imaging (or MRI) devices were introduced in the 1980s in the form of machines manufactured by Technicare, Diasonics, Fonar, and Picker International. The signals detected by these machines were processed by a system that included a microprocessor and a considerable amount of signal-processing circuitry. There were already 140 such machines installed around the world in 1984, and their use has become almost commonplace. Physicians, unconvinced of the reliability of electronic communication equipment and networks, were sometimes reluctant to accept proposals for systems that promised "medicine by wire," but by the end of the century it was becoming apparent that health care was undergoing its own computing revolution.

EMERGING TECHNOLOGIES

The 1990s saw renewed interest in compound semiconductors and hetero-junction bipolar transistors (transistors made of two or more different semi-conductors), using silicon-germanium as well as materials such as such as aluminum gallium arsenide and gallium arsenide. These devices grew out of some of the same research that had led earlier to semiconductor lasers. The semiconductor lasers and LEDs announced in the early 1960s were ineffi-cient and often required very low temperatures to operate. Two researchers, Zhores I. Alferov and Herbert Kroemer, working independently, suggested the use of what they called "heterostructures" to improve them. Junctions created by using two different semiconductors (rather than a "monostruc-ture" of a single type of semiconductor with differently doped regions) gave a different sort of effect and (in theory) allowed designers to control the en-ergy states of electrons and holes. In a heterojunction transistor, for example, the band gaps and other properties of the emitter, base, and collector could be individually shaped, and the characteristics within those structures (such as their conductivity) could even be varied in different parts of the material.

While the fundamental theory of these devices emerged in the early 1960s, it was many years before practical devices appeared. In the mean-time, physicists found ways to improve ordinary semiconductors to create commercial LEDs and lasers. It proved to be extremely difficult to create junctions between dissimilar materials that were free of imperfections in the crystal lattices. The Soviet Union developed heterojunction solar cells using this principle, but other applications had to wait until the late 1980s and 1990s. Part of the reason that other types of devices took so long to de-velop had to do with the cost and complexity of the heterostructures. They added more steps to the production process and put restrictions on how wafers were processed.

IBM struggled for many years to perfect such devices, and then to make them on large-scale integration chips with acceptable manufacturing yield. Researchers finally succeeded by using ultrahigh vacuum chemical vapor deposition techniques developed by IBM engineer Bernard Meyerson. The properties of these devices allowed efficient use at gigahertz (GHz) fre-quencies for analog and digital signal processing, optical communications, radar, and other applications. Research on millimeter-wave integrated cir-cuits, for example, was aimed at producing devices to be used in intelligent cruise control in vehicles. In the mid-1990s, the first commercial ICs em-ploying heterojunction silicon-germanium bipolars also were announced, and this class of devices remains one of several areas of intense interest for electron device engineers.

Another emerging field in the 1990s was magnetic semiconductors, known as "spintronics," developed at Argonne National Laboratory; Motorola; IBM; California Polytechnic State University (Cal Tech); the University of California, San Diego; the U.S. Naval Research Laboratory; and elsewhere. Harkening back to the days of bubble memories, one proposed spintronic device uses ferromagnetic layers sandwiched between spacers and insulators to create a new form of nonvolatile, high-density magnetic memory (MRAM). The field promises devices 100 times smaller than what is being built today, and in recent years spintronics researchers in the United States have received more than $50 million from DARPA, which hopes it will lead to replacements for flash memory chips. The major drawback, however, is the need for near-absolute zero temperatures when operating these devices.

Looking ahead, the U.S. military anticipated that in 10 to 15 years there would be a dramatic shift to spintronic or other "nanoelectronic" circuits with element sizes less than 100 nm. They also forecast a trend toward the integration of electronics, optoelectronics, and microelectromechanical systems (MEMS) on single chips. Nanotechnology, another buzzword of the last decade of the twentieth century, referred to the effort to reduce the scale of chips further than what was possible with photolithography. These devices do not rely on the mass movement of electrons, as in an ordinary transistor. In a nanoelectronic device, devices operate at the atomic or molecular level, and take advantage of so-called quantum effects. By 2000, microelectronics seemed poised on the brink of another era of rapid technological progress, but for the time being the integrated circuit did not appear to have any serious challengers.

EUROPEAN SEMICONDUCTORS EMERGE

European governments, and corporations such as Siemens and Phillips, developed their own semiconductor manufacturing and research facilities in the 1950s, and these were supplemented by foreign-owned plants (such as the several transistor-fabrication plants opened by Texas Instruments in the 1960s). Texas Instruments was, for example, the largest transistor manufacturer in both Great Britain and France by 1968. Similarly, Motorola began operating transistor plants in France. Great Britain, and West Germany in the late 1960s and early 1970s, aggressively marketed transistors in other countries in Europe, grabbing large shares of the market. By the 1990s, spurred in part by the appearance of new products and governmental support for homegrown integrated circuits, several European firms had captured markets for specialty chips and discrete devices. Russia, the Czech Republic, Hungary, and Bulgaria all possessed significant expertise in the integrated circuits area when the Soviet Union collapsed in the 1980s, although engineers in Eastern European countries admitted that many of their designs were exact duplicates of Western designs. Into the 1990s, there were openly published directories cross-referencing Soviet-era and Western chips.

ASIAN SEMICONDUCTOR MANUFACTURERS

Asian firms have been mentioned mainly in passing throughout this book, and this is a reflection on this work's emphasis on cutting-edge technology, the invention of new types of devices, and new production techniques rather than production. From the 1950s, when Sony became the first Japanese transistor licensee, to the end of the 1970s, many of the contributions of Japanese industry were in the area of production. It was the Japanese superiority in manufacturing that allowed them to swallow up much of the market for consumer electronics in the 1960s. The emergence of IC technology impressed upon the leaders of the Japanese manufacturers the technological gap between themselves and their counterparts in the United States in terms of semiconductor manufacturing. In response, Japanese companies negotiated agreements with U.S. firms in order to gain access to IC technology. Toshiba, for example, built a relationship with General Electric, Mitsubishi with TRW, and Hitachi with RCA. Busicom's later calculator line, based on the microprocessor designed by Intel, triggered what became known as the "calculator wars," in which a group of Japanese companies competed fiercely in the field of pocket calculators. By 1978, virtually all of the small and medium-sized companies involved in the calculator wars had gone bankrupt, while larger companies such as Hitachi, Toshiba, Mitsubishi, and NEC had shifted their efforts to producing the memory chips and logic circuits used to make first calculators and then computers. Only Casio and Sharp survived as calculator manufacturers.

Meanwhile, the Sony Corporation negotiated with Texas Instruments from 1963 to 1968 before reaching an agreement that not only gave Sony access to TI's patented integrated circuit technology, but also gave that access to any interested Japanese manufacturer. The five main Japanese companies in the integrated circuit business in the 1980s had gotten their start building computers and IBM-compatible computer "peripherals." Beginning in 1975, the Japanese government sponsored several crash programs that brought the manufacturers together to jointly develop photolithography techniques, optical semiconductors, and high-speed VLSI memory chips. It was in the memory chip field that Asian firms first made their mark. While in the early 1970s the Japanese share of the world memory chip market had been about 5 percent, by 1979, sales of Japanese-made, 16K DRAMs had reached 48 percent. By 1982, the United States had become a net importer of semiconductor products, and its share of the world market had slipped to about 55 percent. Between that time and about 1985, Japanese firms completely overwhelmed U.S. and European competitors in the manufacture of the next-generation 64K and 256K chips, taking over 90 percent of the market for these products. Part of the reason was cost. Competition brought the price of 64K memory chips down to 25 cents, and 256K RAMs were retailing for less than $4.00 in 1985.

One of the areas outside computer chips where Asian firms excelled was in the field of flat-panel displays. After Sharp's early introduction of the LCD calculator in 1975, the company moved into ever-larger LCD displays for personal computers and stand-alone word processors in the 1980s, a 14-inch LCD television in 1988, and then (along with another firm, Seiko) the first generation of LCD projectors beginning in 1989. A second area was in the field of solar cells. Working from technology licensed from RCA in the 1970s, Sanyo improved processes for integrating solar cells using IC production techniques. The Sanyo amorphous silicon solar batteries, announced in the 1980s, were discovered to be more efficient than ordinary solar cells in the presence of fluorescent lighting, opening the door to a lucrative market supplying solar batteries for small devices like calculators, intended to be used indoors. The Canon Corporation also had quite a bit of success using amorphous silicon devices as photoreceptors (part of the copying process) in their line of photocopiers beginning in the late 1980s. A third and perhaps more important field was flat-panel displays. In the 1990s, the firm pioneered in video cameras with small, color LCD viewfinders.

Later in the 1990s, Japanese companies took the lead in other areas of microelectronics, especially RF communications, hybrids, and converter technologies. The high-electron mobility transistor (HEMT), for example, was invented by Takashi Mimura of Fujitsu Corporation in late 1979. Similar in structure to an ordinary MOSFET or MESFET, in the HEMTs the normal n-doped channel is replaced by a junction consisting of two materials of widely differing band gaps. This type of junction creates a thin region where the energy of the electrons is above the conduction band, creating a condition similar to an ordinary channel. The conductivity of the channel can then be modulated by the gate voltage in the ordinary way. Mimura found that the new transistor can operate as a microwave amplifier with an extremely high level of sensitivity. Fujitsu began offering these commercially in 1985. They were used initially for extremely sensitive amplifiers with radio telescopes, but in the 1990s found a major commercial application as part of Direct Broadcast Satellite (DBS) television receivers. Earlier satellite television dish antennas were over 6 feet in diameter, making them suitable only for rural or suburban installations. By combining a new type of satellite with HEMT amplifiers, dish antennas of about a foot in diameter were possible.

U.S. PROTECTIONISM

The entrance of Japanese firms as major producers of memory chips also sparked the first of what became regular boom-and-bust cycles in the chip business. In 1975–1977, NTT funded a VLSI project in which NEC, Hitachi, and Fujitsu participated and that resulted in the development of the first 64K DRAM in 1977. NTT engineers helped Japanese firms overcome technological barriers to commercialization of this product, and by 1981 Japanese firms controlled 70 percent of the world market for 64K memories. Engineers at NTT subsequently developed a 256K DRAM and transferred this technology to four Japanese firms, apparently free of charge. By the mid-1980s, Japanese firms controlled 90 percent of the world market in 256K DRAM.

Japanese firms achieved this remarkable market share in DRAM technology in part by expanding their production capacity far more rapidly than the growth of the market could absorb. In 1981–1982, several Japanese manufacturers flooded the world market with 64K DRAMs sold at prices that were apparently below the cost of production. This occurred during a period of rapid growth in the world memory industry as a whole, and U.S. semiconductor sales alone grew from $8 billion to $14 billion between 1981 and 1984. However, in 1984 the demand for memory chips suddenly dropped off and buyers of memories began to cut back their purchases. Unfortunately, this happened just as U.S. manufacturers were increasing production levels and as Japanese producers were dumping large amounts of their products. This combination was disastrous for U.S. producers. In 1985–1986, the U.S. semiconductor industry suffered $1–2 billion in losses, lost 20 percent of its world market share, and laid off over 27,000 workers. Japanese firms were able to weather the crisis and, in fact, use it to expand their own market share, even though the price of a Japanese 64K DRAM fell from $3.53 in September 1984 to 82 cents in September 1985.

The response in the United States to the rise of Asian semiconductor manufacturing resembled panic. The Defense Science Board of the U.S. Department of Defense published a *Report on Defense Semiconductor Dependency*, pointing out that there were defense-related reasons to protect the semiconductor industry in the United States. The federal government convinced Japan in 1986 to restrain companies from the alleged dumping of memory chips at prices below cost. The laissez faire atmosphere of the industry was weakened further when, in April 1987, the government imposed trade sanctions on Japanese imports. Shortly afterward, Congress discussed spending $500 million over five years to finance a new industrial consortium to be called SEMATECH (for Semiconductor Manufacturing Technology), aimed at collaboratively developing new semiconductor manufacturing technologies. Further, there were widespread accusations of semiconductor "piracy," referring to the common practice of copying successful chip designs by reverse engineering them. Protectionist sentiment heightened, and President Ronald Reagan went so far as to sign into law a semiconductor mask piracy law in 1984.

The pressure was intense on the U.S. manufacturers, many of whom had pioneered in the field. Many, such as MOSTEK in 1985, were forced to close up shop and leave the memory business entirely. Intel stopped making RAM chips to concentrate on other products such as microprocessors. Many others in the industry shifted their focus to manufacturing ASICs, which were then a growth market. Manufacturers, weakened by this competition, were targets for takeover, and the U.S. government stepped in more than once to prevent what it saw as undesirable acquisitions. Fujitsu, for example, was prevented by federal pressure from acquiring Fairchild Semiconductor in 1987.

Several prominent engineers in the IC field participated in an October 1985 seminar on protectionism held on behalf of the U.S. Congress by the IEEE. Gordon Moore, who coined the famous "Moore's Law," presented evidence that Japanese companies were dumping chips in the United States under direction from their government. Michiyuki Uenohara of NEC denied this and added that the Japanese were practicing what had long been considered good business practice in the United States: offering good products at prices that the competition could not match.

With punitive legislation likely, Japanese companies backed away slightly from their earlier competitive stance in 1987. At the same time, the gears of government-industry cooperation began to turn, resulting in the formal establishment of SEMATECH. In earlier times, it might have been considered a violation of antitrust legislation, but the perceived crisis in memory chip manufacturing and a presidential administration friendly to big business helped override those objections. Later, in May 1994, the federal government formed a U.S. Display Consortium along the lines of SEMATECH, headquartered in San Jose, California. It was funded by $1 billion in DARPA money. At the end of the century, these government-industry initiatives had seen some successes, although it appeared to some that new products such as the Internet and the spread of cellular telephony had done more to revive U.S. semiconductor manufacturing. Further, after a boom in the 1980s, there was a general recession in the Japanese economy that, among other factors, led to ongoing weakness in the 1990s. Many Japanese firms moved their production to low-wage areas in Singapore, Malaysia, and Thailand, and concentrated on areas of strength rather than reaching into untested markets.

POWER ELECTRONICS

While vacuum tubes were used almost exclusively for power rectifiers and transmitter applications over a few watts through the 1960s, these applications were challenged as semiconductor power device construction techniques improved. As early as the 1950s in some applications, relatively large semiconductor diodes replaced vacuum tube rectifiers. High-power diodes were supplemented beginning in 1956 by John Moll's invention, the PNPN transistor, which is today usually called a silicon controlled rectifier (SCR). Invented at Bell Labs but commercialized as the silicon thyristor by GE a few years later, this device was a specialized form of transistor that could act as a variable power control. The SCR and a related device, the Triac, had wide applicability in power supplies for various types of equipment, although they were still limited by high cost.

The MOSFET transistor, introduced as a high-speed, low-current switch, became the basis of high-power devices experimentally by the 1970s. Power MOSFETs introduced commercially in 1981 included a device from General Electric that could conduct 60 amps and had a blocking voltage of 600 volts. It was to be used as a low-loss synchronous rectifier in an efficient high-frequency power supply. 200-volt switching and control chips were available by 1983, and it was thought they would soon be challenged by 400-volt chips soon afterward. JFETs made by GE, using a recessed gate structure, could block up to 400v, with a bandwidth of up to 500 MHz. These new devices could be connected directly to the power mains or in appliances plugged directly into a wall outlet. They combined logic with relay or switching functions and often merged MOS and bipolar devices on the same chip. The development of these devices was led by Bell Labs,

Fujitsu, Harris, Hitachi, Motorola, Nippon Electric, NTT, Oki Electric Industry Co., Sharp Electronics, Sprague Electric Co., Texas Instruments, Thompson, and Xerox.

In 1984, more chips were announced that mixed power and logic circuits on the same chip to provide controls for industrial applications such as DC-DC converters, audio amplifiers, DC motor controls, fluorescent light controls, power tools, alarm systems, and home appliances. Thus semiconductors began to replace the transformer in many electrical devices, particularly computers and consumer electronics, by the late 1980s.

Electronic tubes had long been used in electric distribution and transmission systems or other very high-power applications for switching and rectification. It was decades before semiconductor devices could come close to vacuum tube performance, and even in 2000 some types of tubes are still in use. In 1992, for example, vacuum tubes for the switching of power were able to handle 8,500 volts at 3,500 amps with a single device. A "magnicon" tube with 2.6 megawatts (MW) of power and 73 percent conversion efficiency at 1 GHz was developed. Gyratron tubes were then being produced that could handle up to 500 kW at frequencies of up to 110 gigahertz. These frequencies far exceeded the needs of electric power systems, so engineers looked for alternative applications. There was at least one proposal to use such tubes to reduce the half-life of radioactive waste by passing it through a powerful microwave beam. In sum, then, high-power semiconductor devices by the 1980s had taken over many applications where vacuum tubes or transformers had been used before. From power supplies and home appliances to controlling high-voltage transmission lines, power electronics was quietly making strides, though the technology had not become well known to the public.

THE SEMICONDUCTOR LASER IN PRODUCTION

The semiconductor laser, announced in the laboratory years earlier but not widely manufactured, at last found a mass market in the compact disc (CD) audio system. The CD was based on several earlier laser videodisc systems introduced between the late 1970s and early 1980s, all of which used more expensive types of gas lasers. Philips, the sponsor of the first such system, sought to salvage its investment in the failed "DiscoVision" analog optical disc by forming a partnership with Sony to produce a digital audiodisc. The chief obstacle was a reliable semiconductor laser that would operate at the 780-nm wavelength chosen by the designers. For this, Sony worked collaboratively with Sharp Electronics, which had developed a suitable laser in 1981. Sample quantities of these lasers cost $800, but the price had been reduced to $18 by the time the system shipped.

After the compact disc system was first offered in 1983, its sales were also disappointingly slow for the first several years. Prices of players dropped from their initial price of about $2,000 to below $350, and several major record companies committed large parts of their catalogs to the new medium. Sony, the major promoter of the system worldwide, predicted that automotive and portable players would become big sellers, and redesigned their basic laser optics assembly to create a one-third-size model for car players. VLSI circuits handled a variety of functions in the new assembly, including motor speed control, frame sync and error detection, error correction, and data interpolation. Sales of CD systems began to take off by the late 1980s, and would in the 1990s eventually surpass the leading analog technology (which by that time was prerecorded cassettes). Meanwhile, even in the 1980s manufacturers perceived that a major objection to the CD was its lack of recordability. While the first recordable CD system was introduced by Nakamichi USA, it cost $80,000 and was clearly not aimed at the consumer market. Nonetheless, recordable CD products would be reintroduced several times before the end of the century, and sales were rising in the late 1990s. The boost to the laser given by CD and videodisc systems (which returned to prominence in the form of the DVD in the late 1990s) was supplemented by CD-ROM devices for computers, photocopiers, laser printers, and other such systems. The worldwide market for semiconductor diodes reached 100 million devices sold per year by 1995.

Some of this growth by the 1990s was coming from the field of telecommunication, where optical fiber had become a huge growth industry. Optical switches and other types of equipment often used lasers manufactured by a process developed at Rockwell International in the late 1960s. There, Russell Dupuis, Harold Manasevit, and Paul Dapkus had demonstrated an important new process called metal-organic chemical vapor deposition (MOCVD) in 1968. By 1978, Dupuis, Dapkus, and Nick Holonyak were able to use the process to make laser diodes and other devices with very thin layers of single-crystal material—so thin, in fact, that layers of just a single atom's thickness seemed possible (and later were). By the end of the twentieth century, MOCVD was the most widely used process for making LEDs. It could also be used to make solar cells and lasers, and MOCVD lasers would become commercially important in optical communication networks.

OTHER LASERS

The military, scientific, and biomedical markets for lasers that had sustained research in the 1960s and 1970s remained strong in later decades. The Star Wars (SDI) missile defense system conceived by Ronald Reagan was one such military project that had a decisive impact in the field of lasers. While many believed that laser technology could not be used effectively against missiles, Reagan pushed for more research in this area. The idea of a beam-type weapon had existed in science fiction literature since at least 1898, when H. G. Wells wrote of one in *War of the Worlds*. The inventor Nikola Tesla, among others, announced such weapons around the time of World War I, but none was actually constructed and thereafter the idea retreated to comic books and motion pictures. Yet since the first announcement of the laser, engineers had looked seriously at the possibility of using them to fulfill this longstanding dream. In 1985, researchers at Lawrence Livermore laboratories demonstrated a nuclear bomb–pumped X-ray laser and tested another high-energy laser system at the White Sands, New Mexico, proving ground. There, a 3.8-micrometer-wavelength, deuterium-fluoride laser produced a beam with 2.2 megawatts of power. While the *IEEE Spectrum* in January 1986 reported that "it was acknowledged that lasers far brighter would be needed for a strategic defense system," in this demonstration the laser (from a distance that was kept classified) managed to rupture the casing of an old Titan I missile used as a target.

Nicolaas Bloembergen: On the Politics of Star Wars

Nicolaas Bloembergen invented the three-level solid-state maser and is a pioneer in the field of nonlinear optics.

Because of my interest in lasers, I was asked to chair [a] study on directed energy weapons, which I did with Kumar Patel. We were co-chairmen of that study. It attracted a lot of attention, especially in Washington, because the Cold War was still going on, and this was part of Reagan's Strategic Defense Initiative. Even during the Cold War on the basis of our study they cut back their plans on deploying the big weapons in space. . . .

I know that on our committee there were some people on the right and there were people on the left, but we got a report out which was unanimous. There was no minority opinion. We often had long discussions, and then I said, "Look, here we are all engineers and scientists. Why can't we agree on facts?" It was either one of two things. In some cases the question was a purely political one, and in those cases, we said, "That should not go in our report. That is not what we are here for." In other instances, the disagreements were on technical issues. The wording that was used in the draft was not neutral. So then we would go very carefully through it and reword it so that we achieved a unanimous report with people who had a very broad political spectrum of opinion. . . . It may have worked that it helped to bankrupt the Soviet Union, [but] that was a typical question that we wouldn't address. What we were there for was to determine whether it was feasible to design a system of strategic defense by deploying directed energy weapons in space.

Source: Nicolaas Bloembergen, an oral history conducted May 15, 1995, by Andrew Goldstein, IEEE History Center, Rutgers University, New Brunswick, New Jersey.

The dissolution of the Soviet Union that began in the late 1980s seriously undermined the desire to maintain the high levels of defense research spending seen in the early part of the decade. One of the first casualties of reduced spending was Reagan's favored Star Wars program. The X-ray-pumped laser program ended in 1992 without having built a successful weapon, although there were efforts later in the decade to revive aspects of it.

Lasers had a somewhat happier career in medicine. The first neodymium yttrium aluminum garnet (YaG) lasers were approved in the United States in 1985 for medical purposes. The first such device cost $80,000, but later prices came down to the level where smaller organizations or even private practices could afford them. Excimer lasers, for example, were approved for vision correction in 1992 and became one of the most popular forms of laser-based procedure. Plastic surgery, the removal of tattoos, and many other types of operations were possible using lasers in later years.

NEW LIFE FOR THE LED

The LED as a display device had long since been relegated to simple "on-off" indicators and other menial tasks by 1990, but that would soon change. Following the invention of red, yellow, and green LEDs, progress had stalled. Efficient blue LEDs, which would make (through a combination of colors) a white light LED possible, did not emerge until 1994, when S. Nakamura at Nichia Chemical Industries in Japan announced the first practical blue LED. Almost instantly, the LED's importance as a source of light leapt forward, because for the first time a semiconductor light source could compete with Thomas Edison's century-old light bulb. Along with incremental advances in the brightness of LEDs, the devices began to replace incandescent bulbs in applications such as traffic signals and automotive taillights by the end of the century.

OPTOELECTRONICS AND THE FIBER REVOLUTION

One of the most important applications for the laser and the LED in the 1980s and 1990s was a field of engineering known as optoelectronics. Glass fiber conduits for coherent (laser) light were first proposed in 1966 and put into production four years later. Information was transmitted along these optical fibers in the form of pulses of light, representing digital data. After many years of development, optical fiber transmission lines were ready for commercialization. Early in 1980, AT&T requested permission from the Federal Communications Commission to approve a Northeast Corridor optical system running from Boston to Washington, and British engineers began work on a submarine fiber-optic cable.

Such early fiber-optic transmission lines for digital communication were put into service by AT&T and others only for short-haul telephone communication because the relatively large diameter fibers they used were uneconomical for long-range transmission. In an effort to increase the range, Bell Laboratories researchers developed gallium arsenide indium phosphide (GaAsInP) diode lasers operating at 1.3 micrometers. In the meantime, plans continued for medium-length cables and even fiber-optic service to the home: the Canadian telephone company late in 1981 experimentally ran fiber-optic cable to homes in Elie, Manitoba. Although the World Wide Web was still years away, telephone service companies were thinking in terms of all-digital telephony.

Late in 1982, foreshadowing things to come, MCI leased the right-of-way for a fiber line from New York to Washington, D.C., and began constructing it, but AT&T won the race, opening the first New York to Washington, D.C., optical cable in February 1983. This 595-km line used the somewhat less efficient AlGaAs laser and had repeaters spaced 7-km apart. The laying of several fiber-optic land and submarine cables was well underway by 1985 when the first transatlantic fiber-optic system went into service in December 1988.

At each end of a fiber-optic communication cable, light is generated by a semiconductor laser and detected by a semiconductor diode. Yet in addition to requiring suitable fiber, laser, and photodiode improvements, the new lines also highlighted the need for more efficient ways of moving massive amounts of data. What AT&T and others sought was faster "switching" technologies (the computer-like devices that route calls through the network), and preferably switches that did not require the conversion back and forth between optical and electronic signals. Integrated circuits made of gallium arsenide or other compounds from the so-called III-V group (that is, elements from the third and fifth columns of the periodic chart) can operate optically or electronically, and the first commercial switches employing this technology were a natural choice for optical switching. GaAs chips, which a few years earlier had been considered too expensive for any but the military market, and which were heavily subsidized in the development stage by the military, by 1985 were beginning to be used in optical communication. These high-speed circuits were being used for multiplexers (circuits to "piggyback" two or more signals on a single fiber), demultiplexers, repeaters, and other circuits. By 1987, GaAs on silicon was announced, a hybrid of GaAs and ordinary silicon semiconductor technology that opened a doorway to integrated optical and electronic components on a single chip.

By the early 1990s, it was possible to transmit digital data economically for about 135 miles without the need to electronically regenerate the pulses. Such improvements in range paved the way for a greatly expanded communication infrastructure based on fiber optics that came just in time to coincide with the emergence of the Internet as a high-volume, high-speed global data network. The optoelectronics developed for use with fiber-optic systems has in recent years held out as a potential solution for the problem of building faster computers, since these chips could replace the electrical interconnections between chips.

SUPERCONDUCTORS

Some of the newest and most promising materials for the next generation of electron devices are superconductors. The phenomenon of superconductivity was discovered early in the twentieth century, and transistor pioneer John Bardeen earned his second Nobel Prize by developing a theory of superconductivity, but there was little commercial interest in this technology until nearly the end of the twentieth century.

JOSEPHSON JUNCTIONS

Brian Josephson applied superconductor theory in 1962 to suggest the possibility of a device consisting of two superconductors joined by a barrier of nonsuperconducting material. The device could be used as a very fast switch or diode. This led to the experimental development of various devices utilizing the so-called Josephson junctions. Few of these could be used in practical settings, however, because the temperatures needed to achieve superconductivity were so low. The device had its roots in the 1933 discovery by Walter Meissner and Robert Ochsenfeld that the interior of a sample of metal cooled to superconducting temperatures excludes all magnetic fields. This discovery proved that superconductivity involves more than just zero electrical resistance. Other important changes in electrical properties occur during superconductivity as well, one of which is the Josephson effect. According to theory, superconductivity results from the motion of two correlated electrons, termed Cooper pairs, in the superconducting solid. Josephson found that these Cooper pairs move from one superconductor to the other across the insulating barrier under certain circumstances that can be manipulated by the application of magnetic fields. This effect can in turn be used to change the electrical state of the connected superconductors, allowing for the regulation of electricity. Josephson junctions were originally made from lead alloy, but lead was replaced with niobium in 1983 because it was unstable.

The Josephson effect is highly sensitive to electromagnetic fields, and as a result it has many important practical applications. These include the measurement of tiny electrical currents and the detection of weak magnetic fields. The use of superconductors in electronics technology made possible by Josephson junctions has also produced exciting results since the adoption of niobium in 1983, including the development of very sensitive microwave detectors and stable voltage sources. Engineers have also used the devices in digital electronics, where switching times of 9 picoseconds and logic delays of 13 picoseconds have been demonstrated. As a result, they offer the possibility of superfast microcircuits and high-speed computers. Such computers promise to have far greater speed and storage capacity than current technology allows. In addition, since the voltage across a Josephson junction is known through theory to be based on the values of certain constants, Josephson junctions are also used to provide standards for measuring direct-current voltage. Other applications of Josephson junctions have to do with the metrology of high-speed signals and the development of superconducting quantum interference devices (SQUIDs), which are made out of a number of Josephson junctions connected together to form superconducting loops.

New kinds of "high-temperature" superconducting ceramics were discovered in 1987 by Ching-Wu Chu at the University of Houston and Mau-Kuen Wu at the University of Alabama, Huntsville, leading many to speculate that they could form the basis of commercial products. Using yttrium barium copper oxide, these researchers achieved superconductivity at the relatively high temperature of 95 degrees Kelvin (−288°F), which was high enough so that devices made from such superconductors could be kept cold using inexpensive liquid nitrogen. This was a major breakthrough in the field of superconductivity, opening the door to all sorts of possibilities. Within a few years, superconductors with even higher operating temperatures were found, stimulating even more discussion of their possibilities in transmission lines, motors, or even integrated circuits.

Conductus Incorporated in California was one of the first to demonstrate commercial superconducting Josephson junction devices in 1991. Its SQUID magnetometer (a device for measuring extremely weak magnetic fields) employed a high-temperature superconductor of yttrium barium copper oxide. Hypres, Incorporated, of New York also demonstrated a superconducting logic device, this one a 4-bit shift register operating at 9.6 gigahertz and dissipating only 40 microwatts. Operating in liquid helium at 4.2 degrees Kelvin (−452°F), it contained ten thin film layers and thirty-two niobium Josephson junctions on a silicon or gallium arsenide substrate. By 1992, SQUID devices were in use as laboratory standards for the volt and ohm, and the International and Superconductor Technologies Corporation was manufacturing high-performance, superconducting electronic filters for cellular telephone base stations that were widely in use by the late 1990s. The twenty-first century will undoubtedly see more widespread applications of superconductors in device technology.

6

Conclusions

◆

The thousands of engineers who have contributed to the field of electron devices since 1950 may look back proudly on their accomplishments, but the years ahead hold astonishing possibilities. Nowhere is this more evident than in the field of integrated circuits, where these tiny chips consisting of thousands or millions of individual components are themselves being integrated, in a sense, into nearly everything imaginable, up to and including the human body itself.

INTEGRATION AND MINIATURIZATION

This history of electron devices repeatedly returns to certain themes. One of these was the inexorable process of transforming systems made from discrete devices into integrated circuits. The implication of that trend is that following the development of the integrated circuit, most new types of transistors became truly significant only when they could be fabricated as part of integrated circuits. In the immediate future, that will probably remain the case.

Integration has been accompanied by miniaturization, and correspondingly, breakthroughs considered genuinely significant have tended to be those that promised smaller device sizes. There are important exceptions, of course. Devices that cost less to fabricate than competitors are not always smaller in size, and in special cases the performance of a device may outweigh either cost or size. Further, miniaturization may not be particularly important in some fields of engineering, such as power electronics or image displays. Nonetheless, it is clear that miniaturization is an important factor, and one that deserves special attention.

Jack Kilby on the Integrated Circuit

Jack Kilby, working for Texas Instruments, became one of the inventors of the integrated circuit.

I think I thought it would be important for electronics as we knew it then, but that was a much simpler business and electronics was mostly radio and television and the first computers. What we did not appreciate was how much the lower costs would expand the field of electronics into completely different applications that I don't know that anyone had thought of at that time. . . . The real story has been in the cost reduction, which has been much greater than anyone could have anticipated. And it's tremendously broadened the field of electronics. In 1958, a single silicon transistor that was not very good sold for about $10. Today, $10 will buy something over 20 million transistors, an equal number of passive components, and all of the interconnections to make them a useful memory chip. So, the cost decrease has been factors of millions to one. And I'm sure that no one anticipated that.

Source: "An Interview with Jack Kilby," http://www.ti.com/corp/docs/kilbyctr/interview2.shtml.

It is important to remember that miniaturization is a phenomenon with an identifiable beginning and a historical trajectory. It emerged out of specific historical circumstances and was not a necessary outcome of the invention of the transistor or any other device. In the early twenty-first century, the public (or at least the press) seems to treat miniaturization as an inevitability. In fact, one recent study of Moore's Law goes so far as to present a chart that claims to demonstrate that the "devices" (among them telegraph wires) of the nineteenth and early twentieth century also showed a trend toward miniaturization that conformed to Moore's Law. Such an analysis is a gross distortion in that it denies the historical roots of the miniaturization of electronics, which lie in specific events such as the military demands of World War II and the Cold War. At the close of the twentieth century, many of those initial conditions had vanished or changed, yet the future of miniaturization was unquestioned. Moore's Law, much modified in later years to suit the needs of its advocates, had repeatedly been pronounced expired but repeatedly revived. It seemed clear that at some point the miniaturization of integrated circuits would run up against physical or economic limits, but few predicted that miniaturization would slow or stop before those limits were reached. If the history of technology holds any lessons, it is that the "ultimate" form of a technology is never reached, and that physical limits are not the reason why innovation ceases.

GOVERNMENTAL INFLUENCE

Like miniaturization, the influence of governmental and military organizations in the history of electron devices is of central importance. Old economic theories of technological development insisted that all inventions stemmed from basic human needs, and that technological change was evidence of "the survival of the fittest." That snippet from Darwin was even applied to the functioning of an abstract "marketplace," where innovations compete. Historians have shown that such theories are completely inadequate to explain the development of modern technology. Complex social systems that humans maintain, rather than natural laws, prod us to develop or use technologies that have little bearing on essential needs. Further, the free market is largely a myth. In the case of electron device history, it is evident that powerful organizations, such as government agencies, directly or indirectly contributed as much to numerous innovations as simple market forces did. From the birth of the transistor, the military and other government organizations constituted an important market for new device technologies. Moreover, the military's needs, not market competition, led to numerous device developments that might otherwise not have emerged. Some of these devices even became part of systems (such as the DEW line and intercontinental ballistic missiles) that were designed and built but never called into use for their intended purposes. Clearly, the "need" for such technologies was not based on their actual utility but on their potential.

Yet the military-industrial relationship of the Cold War era was hardly a wasted effort. It stimulated device research to a fevered pitch and brought about basic technological breakthroughs that are still significant today. The transistor, integrated circuit, laser, and numerous other technologies all owe a great deal to this governmental intervention.

John Saby on Research Laboratories

John Saby pioneered the junction transistor at General Electric in 1952.

I was in the electronics laboratory [at General Electric in the 1950s]. It was a division laboratory and our function was in between research and development. What we called a division, they now call a group or a strategic business group. With many of the people in the research lab particularly, the higher you go in the hierarchy there, the more they called everybody else in the company peons. But at the working level of research groups there and our research people, they could get back and forth fine. We had a certain amount of freedom to pursue our own ideas. . . . That disappeared later because of changes in management. That was a time in history when it paid to be flexible. I'm not sure it does now; maybe it does. In the early transistor times it did pay to be flexible. But it does always make sense to ask questions of Mother Nature, whose answers are useful. Someone defined pure research as you don't know quite what you're doing and you don't know why you're doing it. Engineering is you know what you're doing and you know why you're doing it. Applied research is somewhere in the middle: you're still doing research, you know why you're doing it, but you may not fully understand what you're doing. You are finding facts of nature, but you know why you're doing it. You're working in that field because you have some faith that the results in that field will be useful to a particular area of the company. Research labs claim not to have that restriction if they're talking to research colleagues in universities, but if you're talking to people in departments, of course, they claim to have it very religiously.

Source: John Saby, an oral history conducted April 10, 2000, by David Morton, IEEE History Center, Rutgers University, New Brunswick, New Jersey.

The research that military needs spurred has waxed and waned, but overall funding for device research fell in the last decades of the twentieth century. Not all of that drop can be attributed to the military. Other factors, such as the deregulation of the telephone and broadcasting industries, certainly undermined the monopolies that sustained basic research efforts, at least in the United States. It is uncertain whether the world will again see levels of basic research that can match those of the period from 1950 to the 1970s, and because of that it is unclear who will discover the next innovation comparable to the transistor or the laser.

INTERNATIONALIZATION

A somewhat different sort of governmental influence that had an important impact on the device field was programmatic support for the manufacturing of electron devices. Beginning in the 1960s, the Japanese government sought to bolster the nation's semiconductor industries in order to increase exports to the United States and Europe. European firms took much the same tack, but did so partly to recapture their own domestic markets. Ironically, the Japanese economy turned downward in the 1990s, leading Japanese firms to shift production to low-wage areas of Asia such as Korea. U.S. and European firms followed suit, and as a result much of the mass production of electron devices by 2000 took place in regions that many in the West probably could not have previously located on a map. Western industry and government leaders, however, have demonstrated an almost schizophrenic attitude toward this "globalization" of the electronics industry. On the one hand, the U.S. government became so concerned about it that it organized a consortium of private firms, known as SEMATECH, in the late 1980s to revitalize semiconductor manufacturing in the United States. On the other hand, U.S., European, and even Japanese firms have willingly and profitably moved production to other countries, a fact that can hardly be interpreted as opposition to dependence on foreign manufacturing. One thing that is certain is that since the 1980s, the dominant position of the United States in advanced device technology has slipped, and many nations now possess strong capabilities in device research and fabrication.

Steward Flaschen on Innovation in Microelectronics

Steward Flaschen was a pioneer in the microelectronics industry at Bell Telephone Laboratories, and later founded TranSwitch Corporation.

Things have totally changed in corporate America. There are no premier research laboratories left in corporate America. almost everyone now has focused research, which you would not call basic research. . . . Now with more focused research and more advanced development, it's coupled very closely to the profits and loss part of the business. There are pluses to that and there are minuses to that. Pluses include that you might get innovation into your products faster. The minus is that you are not going to uncover the new phenomena that lead to new markets ten to fifteen years down the road. That part of it is gone. I think it has moved to biochemistry. I think the DNA research is a beautiful example of the power of the basic research leading to totally new markets fifteen years after the fundamental work is done. I don't see that happening in electronics anymore, while it is happening in biochemistry and biophysics now.

Source: Steward Flaschen, an oral history conducted June 6, 1996, by Frederik Nebeker, IEEE History Center, Rutgers Unviersity, New Brunswick, New Jersey.

UNCERTAINTY

Amidst the uncertainties about the sustainability of Moore's Law, the decline of industrial research and development, and the economic threat of globalization, there were other uncertainties surrounding electronics at the turn of the twenty-first century. Increasingly since World War II, the public had become more conscious of electron devices than ever before. Even though ordinary people experienced these devices as parts of systems, such as televisions, videotape recorders, cell phones, and computers, they began to be more aware of the fact that through miniaturization, these systems were becoming increasingly pervasive. With that pervasiveness came increasing public anxiety. Amid the wide recognition of the microchip as the heart of useful systems such as personal computers was the widespread belief that companies or governments could (or would soon be able to) implant chips in the human body to control the brain. Cellular telephones were embraced by millions of Americans not only for routine communication but also as a safety feature for their automobiles, a lifeline to help in case of an accident, breakdown, or attack. Yet they were also increasingly perceived as the cause of traffic accidents or possibly brain cancer. In a similar way, computers and the Internet were welcomed for their seemingly limitless capacity to inform, entertain, and enhance communication. Yet they were by 2000 also a major source of concern over the distribution of pornography, and a new type of Internet crime, "identity theft," was becoming a major threat. As battery-operated telephones, computers, games, and entertainment devices proliferated, the disposal of their toxin-filled batteries became a major environmental concern, as did the environmental effects of the disposal of millions of obsolete computers each year. The technologies, such as electronic cameras, that had once seemed so promising now increasingly seemed menacing, as the private lives of individuals were recorded in detail by computers or broadcast over the Internet from cameras hidden in walls, disguised as familiar objects, or embedded in cellular telephones.

Like most technologies in history, electron devices helped give people greater control over the uncertainties and shortcomings of their own bodies and lives. Yet the systems into which devices were embedded sometimes came with a high cost in terms of privacy or choice. A famous example was the so-called V-chip. This device relied on established integrated circuit techniques, yet it became the most controversial IC ever designed. It came into existence in the United States following the 1996 Telecommunications Act, which mandated that TV content producers develop a rating system and that receiver manufacturers provide a way for parents to censor what their children could watch. The actual system to implement this was proposed by Canadian engineer Timothy Collings. While it required technical changes in broadcast facilities, it was the "viewer chip" added to the receiver circuit that captured public attention. The chip came under attack from those who saw it as a form of government censorship, because of the potential for abuse of the ratings. Nonetheless, its use was mandated for all sets sold in the United States after January 1, 2000. While the V-chip's impact was still unclear at the time of this writing, it was a chilling demonstration of the way powerful agencies could make top-level decisions about the future of technology, regardless of market or democratic processes.

Y2K

The twentieth century ended with the terrifying prospect that millions of computers and computer-controlled systems might suddenly cease to function at all. The problem was based on the fact that many computers depended on date and year codes generated by software, and traditionally programmers had shortened year designations to just two digits. When '99 became '00, it was predicted that computers would fail because data from 2000 would be identified as having an earlier date than data from 1999. The first suggestion of the problem appeared in the technical press in the late 1970s, and between about 1993 and 2000, governments and military organizations around the world tested their systems and made many expensive changes to them. Operators of power plants, missile systems, accounting department computers, and others around the world found that the "Y2K bug" would in fact cause many failures. There were so many of these systems, however, that they could not all be tested, and there were not always ways to test embedded systems. As the issue rose to the level of popular consciousness in 1998 and 1999, the press fanned the flames, causing many to prepare for disastrous failures of utilities, food supplies, air traffic control, and nearly everything else. While the year 2000 came and went without an apocalyptic disaster, the Y2K problem seemed to heighten the feeling that there was something subtly threatening about the new technology. The engineers who had created these systems and the devices that powered them had little to say, other than to disparage critics as Luddites and pessimists. But what many engineers failed to appreciate was that underlying such criticisms was the issue of public trust; engineers in the late twentieth century lost and failed to recover the status they had once had as society's problem solvers. The Vietnam-era critiques of electrical engineers as technocrats in the service of the establishment, as enemies of the environment, and as socially retarded "nerds" persisted, reinforced in the years after 2000 by the profession's failure to celebrate Y2K as a catastrophe avoided, or even to demonstrate that it was a false alarm. Many of those took from the Y2K experience a feeling of angst, and a general distrust of new technologies and those who created them. It was typical of the early twenty-first century that the pervasive technologies of daily life had become utterly mysterious and at some level, like all things not well understood, ultimately feared.

Glossary

Active. In the context of electron devices, any device that amplifies or switches current.

Audion. Trade name for the first triode vacuum tube. See vacuum tube.

Capacitor. A passive electrical component consisting of two terminals, usually in the form of layers or plates, separated by an insulating layer. In the proper circuit, this device can store an electric charge.

Cathode ray tube. A vacuum tube consisting of an electron gun and a phosphorescent target screen. Electrons striking any part of the screen cause a localized glow. Used as an information display.

Current. Flow of electrons.

Detector. In the context of radio communication, a detector is usually a specialized form of diode, adapted to operation at radio frequencies.

Device. A component or part of an electronic circuit. Examples of devices include diodes, transistors, integrated circuits, resistors, capacitors, and inductors, but usually not interconnecting wires.

Diode. So named because of its two terminals, the diode acts as a one-way valve for current. Can be a vacuum tube or semiconductor device.

Doping. The process of introducing desired impurities to a sample of semiconductor material. This essential step in creating p-n junctions may be accomplished in a wide variety of ways.

Electrode. In the context of vacuum tubes, a metal conductor or terminal inserted through the glass envelope when it is fashioned. A glass-to-metal seal is formed, and electricity can be supplied to components inside the evacuated tube from an external source.

Electron tube. See vacuum tube.

Electronic. A term used since at least 1930 to describe the field of vacuum tubes. Since the late 1940s, the term has also encompassed the field of transistors and related devices.

Field effect device. In current usage, a transistor or related device capable of amplifying or switching in response to an externally applied electrostatic or magnetic field.

Filament. In a vacuum tube, an electrode that is heated, causing it to release electrons.

Fleming valve. A form of vacuum tube diode.

Germanium. A form of semiconductor first used in point-contact diodes. Now rarely used.

Heterojunction devices. Diodes, transistors, or other devices employing two or more differently doped samples of different types of semiconductor crystal.

Incandescent. In the field of lighting, a type of lamp that uses a metal (or other material) element, heated to the point where it glows brightly.

Integrated circuit. A semiconductor device consisting of all or part of an electrical or electronic circuit, fabricated on a single semiconductor sample. In addition to one or more transistors or diodes, the necessary resistors, capacitors, and interconnections are also fabricated using semiconductor materials.

Junction. In semiconductors, a junction is the interface between two differently doped samples of semiconductor crystal.

Junction transistor. The generic name for any transistor in which two junctions are formed between differently doped samples of semiconductor crystal.

Klystron. A form of vaccum tube used to generate microwave radiation.

Laser. Today a word in common use, laser was once an acronym for light amplification by the stimulated emission of radiation. Gasses or semiconductor crystals, bombarded with energy in a certain way, give off light. Carefully designed resonant chambers and mirrors result in the emission of light photons that are all exactly the same frequency, traveling in a tight beam.

LCD. Liquid crystal display. An imaging or display device consisting of a layer of "liquid crystal" material sandwiched between layers of translucent plastic sheets, which may also act as conductors. Liquid crystal materials are a special class of liquids with molecules that form crystal-like structures that can be altered by stimulation from an external electromagnetic field.

LED. Light-emitting diode. All semiconductor diodes emit energy at a particular freuqency, and the LED is made of materials chosen for their ability to emit infrared or visible light. It is similar in principle to a semiconductor laser, but lacks the resonant chamber necessary to create beams of uniform-frequency light.

Logic. In the context of devices, logic refers to the electrical or electronic circuits used to simulate a rules-based decision-making process. These circuits are usually fabricated in the form of semiconductor integrated circuits.

Magnetron. A specialzed form of vacuum tube used to generate microwave frequency radiation.

Maser. Originally an acronym meaning microwave amplification by the stimulated emission of radiation. An electron device, usually a vacuum tube, similar in principle to the laser. See laser.

Microcontroller. An integrated circuit containing most or all of the circuits previously associated with a computer's central processing unit, in addition to most or all of the system memory circuits, input and output circuits, and certain other features. The microcontroller is usually intended to be part of an embedded system rather than part of a standalone microcomputer.

Microprocessor. An integrated circuit containing most or all of the circuits previously associated with a computer's central processing unit.

MOS. Acronym for metal-oxide semiconductor, a type of field-effect transistor consisting of a multilayer semiconductor sample, coated on one or more surfaces with a thin layer of oxidation, upon which is applied a metal (or other material) electrode.

OLED. Organic LED. An LED constructed from so-called organic semiconductors, containing compounds similar to those that make up living systems. Typically, these compounds are in liquid form and are held between two translucent plastic sheets capable of conducting electricity. See LED.

Orthicon. A specialized form of vacuum tube serving as a camera for television.

Passive. In the context of electron devices, any device that carries current but does not amplify or switch it. Passive devices are usually also referred to as "electrical" rather than electronic, although the distinction is somewhat arbitrary.

Piezoelectric. Refers to the property of certain crystalized materials to produce a small electric current when pressed or twisted. Piezoelectric materials including quartz are also commonly used to regulate the frequency of an electric current, since the crystal resonates at a particular frequency.

P-n junction. See junction.

Point-contact devices. Diodes or transistors constructed by touching two or more fine wires to a sample of semiconductor crystal.

Radar. Originally the acronym for radio detection and ranging. Radar is an electronic system using microwave frequency energy to detect, track, or determine the distance of objects such as aircraft.

Rectifier. A diode. The term is usually used to refer to diodes capable of handling high currents.

Relay. A device, once commonly used in telegraphy, which consists of an electromagnetically operated switch.

Resistor. A passive electrical component with poor electrical conductivity. Electricity passed through a resistor dissipates energy as heat. Used in circuits to reduce or regulate voltage.

Selenium. A form of semiconductor known for its sensitivity to light.

Semiconductor. Literally, a substance or compound that is neither a very good conductor nor a very poor one. However, semiconductors are extensively utilized in electronics because of other properties they exhibit. When carefully mixed with small quantities of certain other materials, crystallized semiconductors can be used to switch and amplify electric currents.

Silicon. An extremely common semiconductor widely used in fabricating solar cells, transistors, integrated circuits, and other devices.

Switching. In the context of electron devices, an electronic switch is directly analogous to a mechanical switch, such as a common light switch. In telephone service, the word "switch" usually refers to an entire electronic system that can automatically complete connections between telephone subscribers.

Transistor. Electron device resembling a three-layer sandwich of different semiconductor compounds. Used in circuits to amplify or switch currents.

Triode. A three-element vacuum tube consisting of an electron-emitting filament and a positively charged plate called the anode, separated by a fine wire grid. Used as an amplifier or switch.

Vacuum tube. The generic name given to a class of electron devices consisting of two or more electrical terminals penetrating the walls of an evacuated glass or metal container. The most familiar vacuum tube is the triode. See triode.

Volt. The basic unit of electrical potential. Voltage is roughly analogous to pressure.

Wafer. In electronics, a thin sheet cut from a large, manufactured crystal of semi-conductor material. The wafer is then extensively processed, and multiple integrated circuits are created on its surface. Individual ICs are then cut from the wafer.

Further Reading

Adams, Stephen B., and Orville R. Butler. *Manufacturing the Future: A History of Western Electric.* Cambridge: Cambridge University Press, 1999.

Agrawal, Govind P. *Fiber-Optic Communication Systems.* New York: John Wiley & Sons, 1992.

Alferov, Zhores I. "The History and Future of Semiconductor Heterostructures from the Point of View of a Russian Scientist." *Physica Scripta* T68 (1996): 32–45.

Bassett, Ross. *To the Digital Age: Research Labs, Start-up Companies, and the Rise of MOS Technology.* Baltimore: Johns Hopkins University Press, 2002.

Bertolotti, M. *Masers and Lasers: An Historical Approach.* Bristol, England: Adam Hilger, 1983.

Boot, Henry Albert Howard, and Randall John Turton. "Historical Notes on the Cavity Magnetron." *IEEE Transactions on Electron Devices* ED-23 (July 1976): 724–729.

Braun, Ernest, and Stuart Macdonald. *Revolution in Miniature: The History and Impact of Semiconductor Electronics.* New York: Cambridge University Press, 1978.

Broad, William J. *Teller's War: The Top-Secret Story behind the Star Wars.* New York: Simon & Schuster, 1992.

Bromberg, Joan Lisa. *The Laser in America, 1950–1970.* Cambridge, MA: MIT Press, 1991.

Brown, Ronald. *Lasers: Tools of Modern Technology.* New York: Doubleday, 1968.

Chandler, Alfred, et al. *Inventing the Electronic Century: The Epic Story of the Consumer Electronics and Computer Industries.* New York: Free Press, 1991.

Charschan, S. S., ed. *Lasers in Industry*. New York: Van Nostrand Reinhold, 1972.

Dummer, G. W. A. *Electronic Inventions and Discoveries*, 4th ed. Bristol, England: Institute of Physics Publishing, 1997.

Dupuis, Russell D. "The Diode Laser—The First Thirty Days Forty Years Ago." *LEOS Newsletter* (February 2003). http://www.ieee.org/organizations/pubs/newsletters/leos/feb03/diode.html.

Early, James. "Out to Murray Hill to Play: An Early History of Transistors." *IEEE Transactions on Electron Devices* 48 (November 2001): 2468–2472.

Editors of *Electronics* magazine. *Age of Innovation: The World of Electronics 1930–2000*. New York: McGraw-Hill, 1981.

Esaki, Leo. Quoted in National Forum on Entrepreneurship and Venture Business, "Minutes from the First Meeting of the Board of Directors." March 17, 2000. www.js-venture.jp-eng-03-030-m030_01.html.

Fielding, Raymond. *A Technological History of Motion Pictures and Television*. Berkeley: University of California Press, 1967.

Finn, Bernard, ed. *Exposing Electronics*. Amsterdam: Harwood Academic Publishers, 2000.

Fitzgerald, Richard. "Physics Nobel Prize Honors Roots of Information Age." *Physics Today* 53 (December 2001). http://www.physicstoday.org/pt/vol-53/iss-12/current.html.

"Forgotten Inventor Emerges from Epic Patent Battle with Claim to Laser." *Science* 198 (October 28, 1977): 379.

Fukuta, Masumi. "History of HEMT Transistors." 1999. http://eesof.tm.agilent.com/docs/iccap2002/MDLGBOOK/7DEVICE_MODELING/3TRANSISTORS/0History/HEMTHistory.pdf.

"The Future of the Electron Tube." *IEEE Spectrum* (January 1965): 50.

Granatstein, Victor L., et al. "Vacuum Electronics at the Dawn of the Twenty-First Century." *Proceedings of the IEEE* 87 (May 1999): 702–716.

Gray, George W. "Reminiscences from a Life with Liquid Crystals." *Liquid Crystals* 24 (1998): 5–13.

Gurtel, Fred, ed. "Microprocessors." *IEEE Spectrum* (January 1983): 34–47.

Hecht, Jeff. *Laser Pioneers*. New York: Academic Press, 1992.

Hobday, Michael. *Innovation in East Asia: The Challenge to Japan*. Cheltenham, England: Edward Elgar, 1995.

Hoddeson, Lillian, and Michael Riordan. *Crystal Fire: The Birth of the Information Age*. New York: W. W. Norton, 1997.

Hong, Sungook. *Wireless: From Marconi's Black-Box to the Audion*. Cambridge, MA: MIT Press, 2001.

Howell, Thomas R., et al. *The Microelectronics Race: The Impact of Government Policy on International Competition*. Boulder, CO: Westview Press, 1988.

Husson, S. S., ed. 25th Anniversary Issue *IBM Journal of Research and Development* 25 (September 1981).

IEEE Oral History Collection. IEEE History Center, Rutgers University, New Brunswick, NJ.

Israel, Paul. *Edison: A Life of Invention.* New York: John Wiley and Sons, 1998.

Johnstone, Bob. *We Were Burning: Japanese Entrepreneurs and the Forging of the Electronic Age.* New York: Westview Press, 1998.

Keller, Peter A. *The Cathode-Ray Tube: Technology, History, and Applications.* New York: Palisades Press, 1991.

Kressel, H., H. F. Lockwood, and M. Ettenberg. "Progress in Laser Diodes." *IEEE Spectrum* (May 1973): 59.

Lengyel, B. A. and V. A. Fabrikant. "Evolution of Masers and Lasers." *American Journal of Physics* 34 (1966): 903.

Leslie, Stuart W. "Blue Collar Science: Bringing the Transistor to Life in the Lehigh Valley." *Historical Studies of Physical and Biological Sciences* 32 (2001): 71–113.

Magers, Bernard. *75 Years of Western Electric Tube Manufacturing.* Tempe, AZ: Antique Electronic Supply, 1992.

Meyer, Herbert. *A History of Electricity and Magnetism.* Cambridge, MA: MIT Press, 1971.

Millman, S., ed. *A History of Engineering and Science in the Bell System: Communication Sciences (1925–1980).* Murray Hill, NJ: Bell Telephone Laboratories, 1984.

———. *A History of Engineering and Science in the Bell System: Physical Sciences (1925–1980).* Murray Hill, NJ: Bell Telephone Laboratories, 1985.

Morgan, David P. "A History of Surface Acoustic Wave Devices." *International Journal of High Speed Electronics and Systems* 10 (2000): 553–602.

Morris, P. R. *A History of the World Semiconductor Industry.* London: Peter Peregrinus Ltd., 1990.

Morton, David. *A History of Electronic Entertainment since 1945.* New Brunswick, NJ: IEEE, 1999.

———. *Power: A Survey History of Electric Power Technology since 1945.* New Brunswick, NJ: IEEE, 2000.

Mueller, Charles W. Oral history conducted by Mark Heger and Al Pisky, 1975. New Brunswick, NJ: IEEE History Center, Rutgers University.

Okamura, S. *History of Electron Tubes.* Amsterdam: IOS Press, 1998.

Perry, Tekla S. "Red Hot." *IEEE Spectrum* 4 (June 2003): 26–29.

Reid, T. R. *Chip: How Two Americans Invented the Microchip and Launched a Revolution.* New York: Simon & Schuster, 1984.

Ronzheimer, Stephen P., ed. "A History of Consumer Electronics: Commemorating a Century of Electrical Progress." *IEEE Transactions on Consumer Electronics* CE-30 (May 1984): 11–211.

Seitz, Frederick, and Norman G. Einspruch. *Electronic Genie: The Tangled History of Silicon.* Chicago: University of Illinois Press, 1998.

Shockley, William. *Electrons and Holes in Semiconductors.* New York: Van Nostrand, 1950.

Silicon Genesis Project. "An Oral History of Semiconductor Technology." http://silicongenesis.stanford.edu/complete_listing.html.

Smits, F. M., ed. *A History of Engineering and Science in the Bell System: Electronics Technology (1925–1975)*. Murray Hill, NJ: AT&T Bell Laboratories, 1985.

Snitzer, E. "Perspective and Overview." In *Optical Fiber Lasers and Amplifiers*, edited by P. W. France, 1–13. London: Blackie, 1991.

Stokes, John W. *70 Years of Radio Tubes and Valves*. New York: Vestal Press, 1982.

Torrero, Edward A., ed. "Solid-State Devices." *IEEE Spectrum* (January 1978): 78.

Townes, Charles H. *How the Laser Happened: Adventures of a Scientist*. New York: Oxford University Press, 1999.

Udelson, Joseph H. *The Great Television Race*. University: University of Alabama Press, 1982.

Varian, Dorothy. *The Inventor and the Pilot: Russell and Sigurd Varian*. Palo Alto, CA: Pacific Book Club, 1983.

Volokh, Eugene. *The Semiconductor Industry and Foreign Competition*. Policy Analysis 99. Washington, D.C.: Cato Institute, 1988.

Wolff, M. F. "The Genesis of the Integrated Circuit." *IEEE Spectrum* 13 (August 1976): 45–53.

Index

About the Author

DAVID L. MORTON JR. is a historian of technology with expertise in the history of sound recording, electronics, and electric power. He is the former Research Historian for the Institute of Electrical and Electronics Engineers.

JOSEPH GABRIEL is a doctoral candidate in the department at the State University of New Jersey, Rutgers.